Climate Change as Political Catastrophe

Climate Change as Political Catastrophe

Before Collapse

ROSS MITTIGA

OXFORD
UNIVERSITY PRESS

OXFORD
UNIVERSITY PRESS

Great Clarendon Street, Oxford, OX2 6DP,
United Kingdom

Oxford University Press is a department of the University of Oxford.
It furthers the University's objective of excellence in research, scholarship,
and education by publishing worldwide. Oxford is a registered trade mark of
Oxford University Press in the UK and in certain other countries

Published in the United States of America by Oxford University Press
198 Madison Avenue, New York, NY 10016, United States of America

British Library Cataloguing in Publication Data

Data available

Library of Congress Control Number: 2023946678

ISBN 9780192868879

DOI: 10.1093/oso/9780192868879.001.0001

Printed and bound by
CPI Group (UK) Ltd, Croydon, CR0 4YY

To George Klosko and Stephen K. White,
for abetting my escape from the cave

lucem post nubila reddit

Acknowledgments

This manuscript has benefitted from the encouragement and generous feedback of so many people. I thank everyone who helped me think about the (often quite bleak) issues explored below and apologize to anyone I fail to mention.

Between 2012 and 2018, I had the incredible opportunity to study political theory and ethics with George Klosko, Stephen K. White, Jennifer Rubenstein, Lawrie Balfour, Colin Bird, Dan Devereux, and Talbot Brewer, among others. I thank them for affording me an unusually diverse normative perspective, which has proven particularly valuable in examining an issue as multifaceted as climate change.

In initially developing this project as a graduate student, I worked especially closely with George Klosko, who is the Platonic ideal of a mentor. Along the way, he exhibited a seemingly indefatigable willingness to read every chapter (and several more later left out) multiple times, always pushing me to make more precise, grounded, and charitable arguments. And since departing the University of Virginia (UVa), George has continued unfailingly to provide feedback and encouragement. I owe much to George, both intellectually and professionally. And even though we do not always agree, I am certain that this manuscript, and much of the rest of my work, would not have existed had I not had the good fortune of working with him these last ten years.

Jen Rubenstein's feedback and support has also been invaluable. Jen has consistently challenged me to shore up weak arguments, cut unnecessary ones, and address structural problems in the manuscript. I also benefitted from her deep commitment to democratic politics, especially when the scale of the problem and the consistently disappointing political responses to it have caused me to question mine.

I am grateful to Colin Bird, whose incisive comments (not just on the substance of my arguments but also their presentation), especially during my time at UVa, always pushed me toward better framing and conceptual clarity. Some of the more unconventional ideas I advance here likely would not have made it to print without his encouragement.

Many ideas explored in this manuscript were conceived in conversation with Tal Brewer. I am deeply grateful to Tal for helping me to understand what well-conducted philosophical inquiry looks like (and can achieve). His ability to interrogate contemporary culture and politics without producing what Nabokov would call "topical trash" continues to inspire me.

I owe an enormous debt of gratitude to Stephen White. Were it not for his support (even before he knew me!), I am certain that I would not be a political theorist today. Throughout my time at UVa, and more since graduating, I have benefitted greatly from Stephen's counsel, feedback, and gracious support of my work. I am fortunate to count him as a mentor and friend.

During my time at the Catholic University of Chile's (Pontificia Universidad Católica de Chile) Instituto de Ciencia Política, my colleagues and students have been great sources of insight and support. I especially wish to thank Alfonso Donoso, Tomás Chuaqui, Giovanni Agostinis, Francisco Urdinez, Cristian Pérez Muñoz, Julieta Suarez-Cao, and Catherine Reyes-Hausholder for the conversations we have shared about this project, and their general encouragement. I am also grateful to Agencia Nacional de Investigación y Desarrollo's (ANID) Fondecyt de Iniciación grant program (project no. 11201060) for financially supporting the research contained herein.

In February 2020, just before the global pandemic hit, I presented an earlier version of this manuscript at a workshop hosted by the University of Virginia. That workshop would not have been possible without the financial backing of UVa's Department of Politics (at the behest of John Owen), as well as the personal assistance of George Klosko, Colin Bird, and above all Stephen White—all of whom I thank deeply for their generosity. That the current manuscript differs so dramatically from the version presented there owes to the incisive comments of several outstanding scholars—above all, Brooke Ackerley, Catriona McKinnon, David Morrow, Steve Vanderheiden—for which I am also very grateful.

In much the same vein, I wish to thank Oxford University Press' two anonymous reviewers for their helpful feedback and support, which prompted further rethinkings and revisions, resulting in a tighter, better grounded, and more thematically unified manuscript. I also thank OUP's outstanding acquisitions editor, Dominic Byatt, for supporting the publication of this book and for his patience as I undertook revisions.

Many other friends and colleagues offered insightful feedback along the way. Matthew Adams deserves particular recognition; our innumerable discussions and his extensive feedback have helped me with almost every aspect of this project. Well before the UVa workshop, and many times since, Steve Vanderheiden has provided invaluable guidance. I am also grateful to John Barry, Ali Benli, Simon Caney, Peter Cannavo, Robyn Eckersley, Lisa Ellis, Yogi Hendlin, John Hultgren, Lukas Meyer, Kian Mintz-Woo, Ryan Pevnick, Thomas Pölzler, Antoinette Scherz, and Santiago Truccone-Borgogno for their help on various parts of this manuscript throughout its long development, as well as to Lawrie Balfour, Chris Berk, Sheri Breen, Peter Christoff, Harrison Frye, Constanza Guajardo, Cheryl Hall, Breena Holland, Gregory Koutnik, Jenn Lawrence, Michael Lipscomb, Tim Luke, Jamie Mayerfeld, Calum McGeown, John Meyer, Sasha Mudd, Michael Payne, Glen Peters, Paul Morrow, Romina Rekers, Diego Tapia Riquelme, Andy

Scerri, Christine Winter, and Jingcai Ying. I also benefitted from the comments I received during the 2015 and 2017 Association for Political Theory meetings, the 2016, 2017, 2018, 2019, 2020, and 2021 Western Political Science Association meetings, the 2016, 2017, and 2020 American Political Science Association meetings, and for the audiences at the colloquia and public lectures I participated in with the Instituto de Ética Aplica (PUC), Instituto de Ciencia Política (PUC), WPSA Environmental Political Theory Virtual Community, and the University of Graz.

I am grateful to the many exceptional students I have had since I began to teach in 2012—particularly in the environmental political theory courses I gave at UVa and PUC—for approaching many of the issues this manuscript explores with curiosity, urgency, and determination. Watching many of these students—especially Hannah Beaman, Joyce Cheng, Laura Cross, Gabby Levet, and Matias Cortes Calderon—fight for climate justice alongside local environmental leaders was a powerful antidote to the pessimism and world-weariness that studying climate change all too easily engenders.

Finally, I thank my friends and family. I am especially grateful to my wife, Francisca González, who has been an incredible source of encouragement and solidarity particularly in the moments during which this project seemed most overwhelming, and an acute interlocutor when some of its arguments were still opaque to me. I also thank my suegros, Manolo and Loreto, for their support, especially when it came to allowing me to colonize their living space to compose some of the later parts of the manuscript. Many of the Mittigas—my father Ross, sister Sarah, niece Rylee, and especially my cousin Anthony and my late uncle Mark—have been encouraging throughout. Another cousin, the talented photographer Angela Mittiga, generously allowed me to use her work for the cover of this book, for which I am deeply grateful. I thank my uncle, Bryan Hilton, for preserving my faith in the idea that older generations can see reason and adapt. My friends—Geramie Laufersky, Rob Moyer, W Cook, Nati Leiva, José Caamaño, Seba Jara, Dani Carreño, Pancho Barros, Grace Martinson, Fran Jimenez, Álvaro Rodriguez, Annetta Campaña, Trini Cereceda, Nico Yoma, Manuel Gonzalez, Aaron Jacobs, David Duhalde, Carlos Diaz Cuzmar, and Matías Sepulveda—provided much needed laughs and moral support along the way. Finally, I must mention my cats—the late great Wednesday, Lupe, and above all Isabella—whose warm presence and quiet solidarity made austere mornings spent writing infinitely more tolerable.

Contents

1
Environmentalism, or Barbarism

> In the face of an absolutely unprecedented emergency, society has no choice but to take dramatic action to avert a collapse of civilization. Either we will change our ways and build an entirely new kind of global society, or they will be changed for us.
>
> Gro Harlem Brundtland, James Hansen, and others[1]

In the mountains of northern Uganda resides a remote tribe of about 10,000 people, called the Ik. According to Colin Turnbull,[2] after several years of extreme resource scarcity, the Ik experienced an almost complete breakdown of "family, cooperative sociality, belief, love, [and] hope."[3] As a later commentator elaborated, the Ik came to

> display almost nothing of what could be considered societal organization. They are so highly fragmented that most activities, especially subsistence, are pursued individually ... Sharing is virtually nonexistent. Two siblings or other kin can live side-by-side, one dying of starvation and the other well nourished, without the latter giving the slightest assistance to the other. The members of a conjugal pair forage alone, and do not share food ... Each conjugal compound is stockaded against the others ... Villages have no political functions or organization, not even a central meeting place. Children are minimally cared for by their mothers until age three, and then are put out to fend for themselves. The separation is absolute.[4]

For Turnbull and others, the Ik show us that key social values, like reciprocity, fairness, and mutual aid, "may indeed be basic to human society, but not to humanity."[5]

We need not accept Turnbull's account of the Ik to accept the basic plausibility of his thesis. Political philosophers have long maintained that the principles and rules of justice can only function amidst a backdrop of reasonably favorable material conditions. Faced with extreme scarcity, David Hume famously warned,

[1] Brundtland, et al. 2012.
[2] NB: While influential, Turnbull's study has been the subject of some controversy and criticism (see e.g. Wilson, et al. 1975; Heine 1985).
[3] Turnbull 1972: 289.
[4] Tainter 1989: 17–18.
[5] Turnbull 1972: 289.

Climate Change as Political Catastrophe. Ross Mittiga, Oxford University Press.
 DOI: 10.1093/oso/9780192868879.003.0001

"the strict laws of justice" give way "to the stronger motives of necessity and self-preservation."[6] This idea also stands behind John Rawls's conception of the "objective circumstances of justice," which he defines as "the normal conditions under which human cooperation is both possible and necessary."[7] When severe scarcity forces a society to depart from those normal conditions, Rawls maintains, "fruitful ventures must inevitably break down,"[8] and thus too relations of justice, and eventually social and political stability.

In this book, I refer to a loss of the conditions that make justice and political stability possible as a "political catastrophe." A central claim I will advance is that climate change threatens political catastrophe on a global scale, and potentially within this century. On this view, a key reason for undertaking immediate and aggressive climate action is to preserve the conditions in which fairness, mutual aid, peace, prosperity, and stability are possible.

There are sadly many real-world examples of political catastrophe on a regional scale—that is, of extreme material scarcity precipitating social and political breakdown in particular areas and times. In 2017, for instance, the *New York Times* reported that:

> In the few, long days since [Hurricane] Irma pummeled the northeast Caribbean ... the social fabric has begun to fray in some of the hardest-hit communities. Residents ... spoke about a general disintegration of law and order as survivors struggled in the face of severe food and water shortages.[9]

Or consider the riots that erupted in Haiti in 2008 after months of food scarcity became so extreme that people were forced to eat dried mud cakes.[10] Just a week after these riots began, the Haitian government fell.[11] A year before the Haitian collapse, Mexico was also roiled by hunger. As drought, market speculation, and demand on corn from biofuel producers caused prices to surge, the so-called "tortilla riots" began. Former President Felipe Calderón cautioned that this unrest was just one instance of a larger dynamic: the "problems in Africa, even the Arab spring ... are [all in] some way or another related to the price of food ... [C]ountries should do something and quickly in order to avoid any social and political turmoil around the world."[12]

[6] Hume 1751: sec. III, pt. I.

[7] Rawls 1999: 109. NB: Rawls directly attributes his account to Hume.

[8] Rawls 1999: 110. I set aside consideration of the former state—that of superabundance. For discussion, including doubt about whether justice would become superfluous in such a state, see Estlund 2016.

[9] Ahmed and Semple 2017.

[10] Katz 2008. As one woman explained, eating a mud cake "stops the hunger ... You eat them when you have to" (Carroll 2008).

[11] Delva and Loney 2008.

[12] Jegarajah 2012.

Perhaps nowhere is this dynamic more apparent than in the recent history of Syria, which has been locked in civil war since 2011. Before conflict erupted there, the country experienced a severe, three-year drought—the worst on the instrumental record. Massive water shortages followed, devastating crops and farmed animals. This forced many rural families into already overpopulated urban centers, and caused food prices to soar.[13] Coupled with strident inequality and corrupt government administration, this soon led to conflict. Between 2011 and 2022, up to 610,000 people have been killed,[14] and many millions more permanently displaced.[15]

These phenomena are harbingers of what is to come as the climate crisis deepens. As researchers studying the Syrian case noted, human-caused global warming "increased the probability of severe and persistent droughts in this region, and made the occurrence of a three-year drought as severe as that of 2007–2010 two to three times more likely."[16] Of course, Syria is not uniquely vulnerable. Without drastic cuts in emissions, by 2080, southern Europe, Iraq, Syria, "the breadbasket regions of China," and "some of the most densely populated parts of Australia, Africa, and South America" will all be in a state of "permanent extreme drought," far more severe than the worst periods of the American Dust Bowl. If this happens, "[n]one of these places, which today supply much of the world's food, will be reliable sources of any."[17]

***

These problems have grave moral implications. The inevitable result of extreme and enduring scarcity is hunger and thirst, conflict, suffering, displacement, and death. In this sense, climate change is a moral catastrophe, capable of inflicting a toll greater than anything we have dealt with before as a species.

But a key argument of this book is that climate change is also, or may soon become, *politically* catastrophic: that such extreme scarcity jeopardizes our most important political values, the stability of our institutions and communities, and thus our ability to respond to the very real moral dangers it portends. Once people are no longer able to meet their basic needs without denying others the ability to do the same,[18] the possibility of fair social cooperation, reciprocity, and mutual support vanishes; we enter into a state beyond justice.[19] In its most extreme form,

[13] Kelley, et al. 2015; Welch 2015.
[14] The Syrian Observatory for Human Rights 2022.
[15] Reuters 2017; United Nations High Commissioner for Refugees 2022.
[16] Kelley, et al. 2015.
[17] Wallace-Wells 2017, citing Cook, et al. 2014.
[18] I defend this conception of climate harm, which is basic to the idea of political catastrophe, in Chapter 2, esp. section 2.5.2.
[19] This is not to deny that moral values might continue to guide our conduct, but only that moral action in such dire circumstances would be closer in spirit to supererogation than duty. For a deeper discussion of how the content and application of justice may change in relation to scarcity, see Adams and Mittiga 2021.

political catastrophe devolves into a war of all against all—not Hobbes's state of nature, where collective security and coordination are the major challenges, but something more hellish, in which (even coerced) cooperation is unable to overcome scarcity, and thus satisfying one's basic needs necessitates conflict. Should such extreme conditions endure, collapse[20] will inevitably follow.[21]

This not to say that justice has no relationship to political catastrophe. Quite the opposite: a central claim of this book is that each generation has a duty to take all reasonable measures to prevent a loss of the conditions that make justice possible—that is, to avoid political catastrophe.[22] On this view, we commit a grave injustice whenever our unnecessary actions or inexcusable omissions[23] risk forcing others into conditions so bleak that the normal principles and rules of justice no longer apply. We have a political and moral obligation to prevent this from coming to pass—to not cause harm, and more strongly to protect those who are most vulnerable to such climate-induced harm, including future people. So far, we have failed at this.

Beyond compromising justice, extreme scarcity fractures social cohesion and weakens institutions, thus imperiling political legitimacy and stability, both within states and internationally. These strains make responding effectively to climate change even more difficult and therefore less likely to succeed. In many cases, the problem of scarcity compounds itself in a political feedback loop.[24] Once crops fail, pressures on governments increase, which (in many cases) leads to greater conflict, more erratic policies, and institutional failure, which in turn further disrupts agriculture and water management, lowering harvest yields and increasing prices and deprivation even more, leading to even greater conflict, and so on.[25]

Here again Syria is a case in point. Since war broke out, much of the country's agricultural infrastructure has been destroyed, and with it the government's ability to distribute seeds, buy crops, and regulate prices has been critically weakened.[26] Moreover, because of security issues, the Syrian farmers who still are able to produce crops are often unable to bring them to market. The end result of this turmoil is obvious in the numbers. Whereas before the war Syria produced nearly 4 million tons of wheat annually (consuming about 2.5 million domestically, and exporting

[20] The conceptual focus of this book is on catastrophe rather than collapse. Nevertheless, it would be helpful to indicate what I have in mind when I refer to the latter. In the most general terms, I understand collapse to refer to a disintegration of collective capacity and/or order, in excess of what can be remedied or reversed by the existing, countervailing forces of repair, replenishment, and regeneration. Along with several co-authors, I articulate and defend a similar conception in Steel, et al. unpublished.

[21] There are empirical studies of precisely this phenomenon. See, for instance, Tainter 1989; Butzer and Endfield 2012; Middleton 2017; Haldon, et al. 2018; Richards, et al. 2021; or, for a more accessible treatment, Diamond 2004.

[22] This claim is developed and defended in Chapter 2.

[23] See Chapter 2, section 2.5.2 for a discussion about what counts as "necessary" and "unnecessary" in this context.

[24] See e.g. Richards, et al. 2021.

[25] World Bank Group 2016.

[26] Peel and Blas 2013; Bulos 2016; Dahan 2016.

the rest), in 2016 farmers there produced just over 450,000 tons—far short of the amount necessary to feed even the government-held portions of the country.[27]

Syria is not unique in this regard. Every year the Fund for Peace produces a well-known ranking of countries according to "their vulnerability to violent internal conflict and societal deterioration."[28] This "Fragile States Index"[29] is based on twelve economic, political, social, and military (or "cohesion") indicators.[30] For each indicator, states are assigned a score of zero to ten, with higher scores indicating "heightened vulnerability of collapse and conflict." Alarmingly, in the 2021 report, thirty-one states had total scores in excess of 90. And of those thirty-one, six were above 100 (putting them in the "high alert" category), and three above 110 (placing them in the relatively new "very high alert" category).[31] Critically, five of the nine states topping the fragility index also occupied the highest positions in the 2021 Global Hunger Index report, and all were among the top 20.[32] Of course, this is not to suggest that states always or exclusively fail because of scarcity, or even that scarcity invariably causes state failure (rather than, say, state failure causing scarcity). The point of identifying this correlation is simply to suggest that extreme scarcity and political instability often go hand in hand and can certainly compound one another, thereby crippling an effective response to the morally catastrophic elements of material deprivation.[33]

No matter how we delineate this dynamic, climate change is only likely to exacerbate it. In a widely discussed 2012 report, the World Bank warned that, at our current rate of emissions, "we're on track for a 4°C warmer world [by 2100] marked by extreme heat waves, declining global food stocks, loss of ecosystems and biodiversity, and life-threatening sea level rise." In such a world, the report stresses, there is simply "no certainty that adaptation ... is possible."[34] Of course, since 2012, many states have committed themselves to taking greater action—most substantially, in line with their proposed "nationally determined contributions" to the 2015 Paris Agreement. Yet, as the latest Intergovernmental Panel on Climate Change (IPCC) Reports (AR6) show, even assuming the Paris Agreement pledges were fully realized—which, given current action deficits, is far from certain to be the case[35]—they would still fall considerably short of what is needed to limit warming

[27] Dahan 2016.
[28] Brown 2009: 20. The Organisation for Economic Co-operation and Development (OECD) and World Bank also produce fragility indices, though unfortunately both maintain them behind paywalls.
[29] Previously it was referred to as the "Failed States Index."
[30] For a description of these indicators, see The Fund for Peace 2021: 40.
[31] The Fund for Peace 2022.
[32] von Grebmer, et al. 2021: 13.
[33] In this sense, it may be helpful to think of a concern over political catastrophe as a second-order concern; we want to preserve states and order to preserve our ability to respond to (and mitigate) the moral effects of extreme scarcity, which is our primary or first-order concern. I am grateful to George Klosko for this point.
[34] World Bank 2012a; World Bank 2012b: esp. xvii-xviii. See also Klein 2014: 13.
[35] Dennis 2021; Plumer and Popovich 2021.

to 1.5°C, and are "likely" to overshoot 2°C.[36] In fact, on all but the lowest emissions scenarios, the IPCC currently predicts median warming of between 2.7° and 4.4°C by 2100[37]—which, even at the low end, would be enough to precipitate ecological disaster in many parts of the world.[38] And there is actually some evidence that these are overly conservative estimates of our current warming trajectory.[39] In any case, the World Bank report's essential claim that climate change presents a grave risk to food security, water access, biodiversity, and the other fundaments of life on this planet has been supported by many subsequent studies focused on precisely these issues,[40] including the United Nations Food and Agriculture Organization's (FAO) 2016 report, *Climate Change and Food Security: Risks and Responses*,[41] and the IPCC's 2019 Special Report on *Climate Change and Land*.[42]

If this is right—if we are credibly at risk of a much hotter world, to which adaptation would be extremely difficult or impossible—then bare assertions of power and violence may offer the only means for attaining essential resources, before collapse. In this sense, climate change represents a major threat to global security. There is now a rapidly expanding literature on the relationship between climate change and conflict. Some have argued that, for every 0.5°C of warming, armed conflict is 10–20 percent more likely to occur.[43] The reasons for this are manifold, but a (perhaps *the*) central factor is, again, resource insecurity. As Stanford climate researcher Marshall Burke explains:

> Hot temperatures reduce agricultural productivity, lower crop yields, and ... for farmers who are close to subsistence, this could alter their incentives to start or join a conflict ... They need to put food on the table. And joining a conflict is one way to do that.[44]

These kinds of conflicts all too often manifest in ethnic violence. In 1998, for instance, following serious long-term food shortages and mass unemployment, riots erupted throughout Indonesia, resulting in up to 1,000 deaths, many dozens of reported rapes, and hundreds of millions of dollars in property damage[45]—the

[36] IPCC 2022b: 18–19; see also Liu and Raftery 2021.

[37] With the median warming estimated to be about 2.4°C by 2100 (IPCC, AR6, WGI, SPM, p. 18).

[38] Wunderling, et al. 2023.

[39] Brysse, et al. 2013; Oreskes, et al. 2019.

[40] See especially Mehrabi 2020; Kornhuber, et al. 2023; Richards, et al. 2023. The valuable work being done on existential risk and climate change is also highly relevant here (see e.g. Bostrom 2013; Kareiva and Carranza 2018; Xu, et al. 2020; Beard, et al. 2021; Bradshaw, et al. 2021;Huggel, et al. 2022; Kemp, et al. 2022; Ripple, et al. 2022). For a helpful overview of this emerging literature, see Steel, et al. 2022.

[41] Gitz, et al. 2016.

[42] IPCC 2018a.

[43] Burke, et al. 2015; Carleton, et al. 2016; both cited in Wallace-Wells 2017. In some cases, temperature increases alone are enough to spur aggression. According to one report, we could see as much as a 54 percent increase in armed conflict in sub-Saharan Africa by 2030 (Burke, et al. 2009).

[44] Quoted in Wallace-Wells 2017: fn.108.

[45] BBC News 1998; Purdey 2006.

great majority of which were inflicted on Indonesia's ethnic Chinese population,[46] which, of course, played no role in precipitating the food crisis.

Indonesia is not unique in this regard. In an illuminating book, Timothy Snyder documents the food security issues and "ecological panic" Germany experienced in the early twentieth century, and how that fueled the Nazi party's ascension and ultimate aggression. On Snyder's telling, Hitler obsessed over the idea that Germans could not be fed "from their own land and territory" and used this to justify his *Lebensraum* policies.[47] Joseph Goebbels certainly espoused such a view, arguing publicly that a war of territorial expansion and mass extermination was necessary to ensure "a big breakfast, a big lunch, and a big dinner" for all Germans.[48] Snyder finds the same logic in more recent ethnic-fueled atrocities:

> The mass murder of at least 500,000 Rwandans in 1994 followed a decline in agricultural production for several years before. Hutus killed Tutsis not only out of ethnic hatred, but to take their land, as many genocidaires later admitted. In Sudan, drought drove Arabs into the lands of African pastoralists in 2003. The Sudanese government sided with the Arabs and pursued a policy of eliminating the Zaghawa, Masalit and Fur peoples in Darfur and surrounding regions.[49]

He takes this as evidence that "contemporary environmental stresses could encourage new variations on Hitler's ideas, especially in countries anxious about feeding their growing populations or maintaining a rising standard of living."[50] Climate change is the largest threat in this regard—a key driver of a new ecological panic.

Even without manufacturing or exploiting ethnic hatred, it is all too easy to imagine how insecure regimes could exploit fear over *actual* water and food scarcity to justify extreme authoritarian policies: for example, tightly restricting immigration (and emigration), militarizing police forces, increasing detention rates, or denying essential resources to refugees, prisoners, or the poor. Europe offers a case in point. As millions of Syrians have fled there, a new jingoistic politics has emerged, with far-right parties advocating much stricter (and often crueler) forms of border enforcement, immigrant quotas and deportations, and, in some cases, the resumption of full national sovereignty, unimpeded by Brussels.

The potential dissolution of the European Union—an institution founded on the aspiration of continental peace and prosperity—suggests the extent to which climate-induced resource scarcity could usher in a new era of international

46 Panggabean and Smith 2011.
47 Snyder 2015b. For his extended account, see Snyder 2015a.
48 Snyder 2015b.
49 Snyder 2015b.
50 Snyder 2015b.

brinkmanship and interstate conflict.[51] This at least appears to be the view of various security agencies and officials. Consider the US Department of Defense's 2014 Quadrennial Defense Review:

> Climate change may exacerbate water scarcity and lead to sharp increases in food costs. The pressures caused by climate change will influence resource competition while placing additional burdens on economies, societies, and governance institutions around the world. These effects are threat multipliers that will aggravate stressors abroad such as poverty, environmental degradation, political instability, and social tensions—conditions that can enable terrorist activity and other forms of violence.[52]

In a report to the US Senate in 2013, then Director of National Intelligence James Clapper similarly emphasized that "extreme weather events" of the sort caused by climate change "will increasingly disrupt food and energy markets, exacerbating state weakness, forcing human migrations, and triggering riots, civil disobedience, and vandalism."[53] Perhaps even more strikingly, when asked what he regarded as the biggest global security threat, Navy Admiral Samuel Locklear—who was "America's top military officer in charge of monitoring hostile actions by North Korea, escalating tensions between China and Japan, and a spike in computer attacks traced to China"—cited climate change.[54] In recent years, this view has become increasingly mainstream, particularly in the defense community. The White House's latest *National Security Strategy* report, for instance, states plainly that:

> Of all of the shared problems we face, climate change is the greatest and potentially existential for all nations. Without immediate global action during this crucial decade... [t]ensions will further intensify as countries compete for resources and energy advantage—increasing humanitarian need, food insecurity and health threats, as well as the potential for instability, conflict, and mass migration.[55]

[51] Notably, Peter Gleick (1991) predicted precisely this kind of conflict as the Cold War concluded.

[52] US Department of Defense 2014b: 8. In another report, the Department of Defense warns that "Rising global temperatures, changing precipitation patterns, climbing sea levels, and more extreme weather events will intensify the challenges of global instability, hunger, poverty, and conflict. They will lead to food and water shortages, pandemic disease, disputes over refugees and resources, and destruction by natural disasters in regions across the globe" (US Department of Defense 2014a, cited in Scranton 2015: 15). Notably, NATO also refers to climate change as a "threat multiplier" (Nuccitelli 2017).

[53] Scranton 2015: 15.

[54] Bender 2013.

[55] The White House 2022: 9.

Those who stand to gain have also taken note of the potential for resource-based conflict stemming from climate change. A 2013 report from the global weapons giant, Raytheon, observes that "climate change may cause humanitarian disasters, contribute to political violence, and undermine weak governments," before disturbingly concluding that "[e]xpanded business opportunities are likely to arise as a result of these security concerns and the possible consequences."[56] Although robust quantitative research on the relationship between conflict and climate change is still relatively inchoate,[57] existing evidence suggests that the frequency of "intergroup violence" could increase by as much as 50 percent by 2050.[58]

While it is impossible to predict precisely how these new climate-resource conflicts or wars may unfold, scholars have already drawn attention to one plausible route.[59] Affluent food-importing nations have begun purchasing or leasing huge tracts of arable land and freshwater rights, often in countries that have high rates of domestic food insecurity and hunger.[60] For instance, by 2009, China had purchased rights to nearly 2.8 million hectares (6.9 million acres) of arable land in the Democratic Republic of Congo (DRC).[61] The DRC nearly tops the Fragile States Index (which, again, measures vulnerability to political collapse), and has a poverty rate of nearly 64 percent.[62] Moreover, from 1998 to 2007, an estimated 5.4 million DRC citizens died mostly through starvation (as conflict severely disrupted agriculture), and to this day the country relies heavily on food aid. China is not alone in participating in what some are calling "land-" and "water-grabbing."[63] Many states—including Russia, Saudi Arabia, the US, the UK, Qatar, the UAE, South Korea, Kuwait, India, Egypt, Japan, Jordan, and others—and even some private corporations are appropriating arable land and potable water in Ethiopia, Sudan, Turkey, Ukraine, Kazakhstan, the Philippines, Vietnam, Brazil, Angola, Congo, Zambia, Myanmar, and Mozambique, among other places.[64] By and large, the purchasers are affluent states dependent on food imports, while many of the host countries are, like the DRC, politically unstable, poor, and, ironically, dependent on foreign food assistance.[65]

Among the many concerns land- and water-grabbing raises, chief is the potential for neocolonial assertions of power and outright conflict. In some places already, foreign security forces have been installed to protect their home governments' land purchases from displaced farmers and hungry citizens.[66] It is also hard not

[56] Schulman 2013. See also Klein 2014: 9.
[57] Buhaug, et al. 2023.
[58] Hsiang, et al. 2013.
[59] See e.g. Aton 2017.
[60] Brown 2009: 10; Swanson 2015; Johansson, et al. 2016.
[61] Brown 2009: 11.
[62] World Bank 2022; World Food Programme 2023.
[63] Rulli, et al. 2013.
[64] Brown 2009: 10–11; Swanson 2015. See also Rulli, et al. 2013; Seaquist, et al. 2014.
[65] Brown 2009: 10–11. There are of course exceptions.
[66] Brown 2009: 12.

to interpret Russia's invasion of Ukraine—and in particular its attempted illegal annexation of the extremely fertile Donbas region—as at least partly motivated by the desire to obtain greater food security.[67] As future climate shocks further constrain agricultural output globally, this dynamic will almost certainly become more widespread and aggressive. At what point will export zones simply become "fortified colonies?"[68] When will states begin to assert, through sheer military might, their "right" to the resources they purchased or leased? At what point will they topple the (often unstable) regimes whose resources they depend upon and install new leaders or institutions that are more amenable to such a colonial relationship? It is impossible to answer these questions now, and we may hope that we never have to. But considering that virtually every powerful country on earth—including every permanent member of the United Nations (UN) Security Council—now possesses land or water rights in other countries, a future of resource colonialism and conflict is hauntingly plausible.

***

Though in some ways the future world I am describing is not far off, it has yet to fully arrive. We *can* still avert politically catastrophic climate change. The subtitle of this book—*Before Collapse*—is meant to express this radical hope.[69] A "radical" hope, as I understand it, is one detached from optimism.[70] This is appropriate in the present context. The world has known about the causes and potential consequences of climate change for more than three decades, and yet, in most countries political leaders of every ideological stripe have failed to take anything but the most minimal and hollowly symbolic actions. Since the first international climate change treaty was signed in 1992—which formally committed states "to protect the climate system for present and future generations"—annual emissions have *increased* by more than 50 percent.[71] Moreover, as the IPCC observes, "[a]verage annual GHG emissions during 2010–2019 were higher than in any previous decade."[72] Consequently, while not yet inevitable, the likelihood of climate catastrophe occurring is real and growing every day. There is now scant time left to prevent crises of unimaginable magnitude—crises of thirst, hunger, neofascism, colonial domination, war, and ultimately moral-political collapse—from which we may never recover.

[67] After all, Ukraine has suffered under some form of colonial food politics for many centuries (Snyder 2022).

[68] Snyder 2015b.

[69] I thank Michael Payne for this point.

[70] A surprising amount of literature explores this conception of hope, or one very much like it (see e.g. Lear 2006; Williston 2012; Martin 2013; Eagleton 2015; Williston 2015: ch. 6; Roser 2020; Moellendorf 2022); I thank one of the book's anonymous reviewers for pointing me to some of this work. I elaborate the idea of "radical hope" in the context of climate disobedience in Chapter 4, section 4.5.2.

[71] As of 2019 (Ritchie, et al. 2020).

[72] IPCC 2022b: 10.

Apart perhaps from the specter of nuclear annihilation, human civilization has never had to reckon with a threat so encompassing and final as that of climate catastrophe.[73] Taking this threat seriously serves as something of a stress test for political theory: it reveals how some of the most common ways we think about politics fall short or break down.[74] This basic insight informs my approach throughout the book. Much as Michael Walzer once argued that "supreme necessity" alters the contours of what is permissible in war,[75] I start from the premise that the possibility of politically catastrophic climate change upends many of the most basic and widely shared assumptions in liberal and democratic thought, including those concerning justice, legitimacy, emergency power, and disobedience.

This first becomes clear in Chapter 2, which centers on the following question: assuming the fairest distribution of burdens will not always be the most effective for achieving an important political end, which value should take priority? Contemporary political theory routinely prioritizes fairness—so much so that many would sooner abandon cooperation before accepting an unfair allocation of benefits and burdens. But this view becomes harder to sustain in the context of the climate crisis; for, after decades of inaction, we may simply no longer be able to prevent catastrophic outcomes without imposing some unfair burdens, costs, or developmental delays.

Under what conditions, if any, would sacrificing fairness be justifiable? The basic argument of Chapter 2 is that subordinating fairness to efficacy becomes permissible when the material conditions that make fairness possible are themselves are at risk—i.e. whenever we are faced with (a credible threat of) political catastrophe. This is highly relevant to the case of climate change, which, if left unchecked, promises scarcity extreme and enduring enough to render fair social cooperation, reciprocity, mutual aid, and political stability impossible. A central claim of Chapter 2, then, is that we have a duty to prevent this dreary outcome from coming to pass—that is, to undertake all measures needed to avert politically catastrophic climate change, again, even if doing so must come at the cost of fairness. This, of course, is not to suggest that fairness has no value or that we should not strive to act in as fair a way as possible, but only that we must not make the implementation of effective precautionary action against credible threats of political catastrophe *conditional* on satisfying fairness standards or principles.

This is a politically significant point, as objections to climate action are often tendered in the language of fairness and rights. For instance, in defending its first, highly unambitious pledge to the Paris Agreement, India—the world's third largest

[73] George Kateb (2002) once argued that the advent of nuclear technology so profoundly altered our political reality that it called into question the adequacy of the political theory canon, as it existed up to that point. I believe the same could be said with respect to the climate crisis.

[74] Several political theorists have recognized this—though perhaps none more than Dale Jamieson (e.g. Jamieson 1992, 2014).

[75] Walzer 2006: ch. 16.

emitter—argued that Indians had a moral "right to grow," and furthermore, given that they had contributed very little to the problem, at least *on certain metrics*, demanding more action of them would be unfair.[76] Highly developed states have made similar—if clearly less compelling—appeals to fairness when attempting to justify their own inaction. Consider former President Donald Trump's remarks as he attempted to withdraw the US from the Paris Accord:

> The bottom line is that the Paris Accord is very unfair, at the highest level, to the United States ... I will work to ensure that America remains the world's leader on environmental issues, but under a framework that is fair and where the burdens and responsibilities are equally shared among the many nations all around the world.[77]

While we may not be surprised to find political leaders exploiting the language of justice to defend inaction, strikingly, several influential economists, like William Nordhaus, have argued against undertaking strong mitigation measures on the very same grounds. They insist that, because future generations are likely to be considerably wealthier and more technologically advanced, requiring those of us alive today to undertake expensive climate action would amount to unfairly expecting poorer people to sacrifice for the benefit of richer ones.[78]

The problem here is not just disagreement about how to define fairness. For even if it were possible to coalesce around a single conception, we may simply have no choice but to violate it if we are to avert (or at least limit the risk of) climate catastrophe. This is particularly clear in the intergenerational context: unless those of us alive today are willing to make potentially substantial sacrifices for the sake of our (perhaps far richer) descendants, then we risk leaving them to contend with a calamity that no amount of capital can alleviate.[79] But the same may also hold *intra*generationally, that is, among states today: there may be no way of undertaking effective precautionary action without, for example, imposing unfair

[76] Specifically, in the Orwellian-titled section, "Considerations of Fairness and Ambition," the Indian government argued: "Both in terms of cumulative global emissions (only 3%) and per capita emission (1.56 tCO_2e in 2010), India's contribution to the problem of climate change is limited but its actions are fair and ambitious."

[77] Trump 2017.

[78] See e.g. Lind 1995: 384; Nordhaus 1997, 2008a, 2008b; and, for discussion, Rendall 2011: 886.

[79] This has been called the "tragedy of the time horizon"—a phrase that some attribute to Mark Carney, former Governor of the Bank of England, from a speech he gave in 2015 (Frame and Cradock-Henry 2022). Kim Stanley Robinson (2020: 173) offers a helpfully lucid description of the concept: "What we do now creates damage that hits decades later, so we don't charge ourselves for it, and the standard approach has been that future generations will be richer and stronger than us, and they'll find solutions to their problems. But by the time they get here, these problems will have become too big to solve. That's the tragedy of the time horizon, that we don't look more than a few years ahead, or even in many cases, as with high-speed trading, a few micro-seconds ahead. And the tragedy of the time horizon is a true tragedy, because many of the worst climate impacts will be irreversible. Extinctions and ocean warming can't be fixed no matter how much money future people have, so economics as practiced misses a fundamental aspect of reality."

costs or developmental delays on relatively poorer states like India or China, for which they may never receive full compensation.

This should not be understood as an assertion of realpolitik. For, unlike most realist accounts, Chapter 2's justification of morally exceptional action is not predicated on a rejection of the overwhelming value of justice (in crisis situations or otherwise), but is rather an affirmation of it. My central argument is, again, that we may only relax or suspend adherence to the principles and rules of justice *for the sake of justice itself*—that is, for the sake of preserving or restoring the conditions that make justice possible. In this way, the concept of political catastrophe gives content to the idea of "necessity,"[80] particularly as it is used to justify states of exception and emergency power.

Invocations of necessity are already shaping climate politics. Consider that, as of September 2022, eighteen national governments, the EU, and thousands of municipalities—together representing more than 1 billion people—have declared a state of "climate emergency."[81] Emergency situations, of course, call for emergency power, which can often be authoritarian in character and scope. Perhaps this explains the growing attraction to "authoritarian environmentalism":[82] for, while most would agree that authoritarian governance is bad and something to be avoided, it also appears increasingly likely that undertaking climate action at the speed and scale needed will require occasionally defying democratic decisions or liberal constitutional restraints. While some may wish to resist the idea that liberal democracy poses an impediment to climate action, consider how often (even low-cost) policies, like marginal carbon taxes, have been defeated at the ballot box or through protest; how constitutional guarantees of free expression have left little legal recourse for combatting well-funded and highly effective climate disinformation campaigns; or how attempts to regulate even modestly the most heavily polluting industries have often been frustrated by entrenched notions of property rights. Should this continue, preventing politically catastrophic climate change may necessitate relaxing or suspending certain liberal-democratic constraints, and thus embracing authoritarian emergency power.

Could such authoritarian power ever be legitimate? This question comprises the focus of Chapter 3. There, I argue that ensuring basic safety and security—which means satisfying the core precautionary duties laid out in Chapter 2—is an essential, if often overlooked, aspect of political legitimacy. According to this view, governments that are able but unwilling to protect their citizens from (credible threats of) political catastrophe are, for that reason and to that extent, illegitimate.

80 See Chapter 2, section 2.4.1.

81 Included here are Canada, Argentina, Spain, Austria, France, Malta, Bangladesh, Italy, Andorra, the Maldives, South Korea, Japan, New Zealand, Singapore, Fiji, Peru, and Vanuatu (Climate Emergency Declaration 2022).

82 The terminology varies between authors. Some refer to "authoritarian environmentalism," others to "environmental authoritarianism," "eco-authoritarianism," or, less commonly, "green authoritarianism." See e.g. Beeson 2010; Gilley 2012; Han 2015, 2017.

Although this security-centric conception of legitimacy may strike some as troubling, it clearly comports with ordinary political practice. In times of war, for instance, authoritarian impositions of power, including those that curtail democratic processes or basic rights, are often thought legitimate to the extent that they are necessary to protect citizens and restore normal conditions. Likewise—as those who have survived COVID-19 can attest—during a health emergency, severe and enduring limitations to rights to free movement, association, and speech can become legitimate techniques of government, even in robustly liberal-democratic states. This suggests that in crisis moments political legitimacy may not only be compatible with authoritarian governance but *require* it. Conversely, stringent adherence to liberal-democratic values may diminish legitimacy insofar as it inhibits responding effectively to grave security threats, again like that posed by climate change. Of course, this should not be taken as an endorsement of authoritarianism, but rather as a *warning*: that should we wish to avoid legitimating authoritarian power, we must do all we can to prevent emergencies from arising that can only be resolved by such means.

Adequate action remains forthcoming, however. And now, after decades of dilatory politics, we are closer than ever to collapse. In this sense, governments the world over face a serious, and deepening, legitimacy crisis—above all, the great historical and contemporary polluters, like the US, China, India, the EU, the UK, Brazil, Russia, Japan, Saudi Arabia, and a handful of others. For they, more than all others, are responsible for delivering us to this dire point, and may alone be capable of preventing the situation from becoming even worse. Every day that they fail to take sufficient action only further erodes their legitimacy.

Illegitimacy, of course, often inspires popular resistance, and the climate crisis is no exception. In recent years, there has been an explosion of climate activism, including innovative acts of civil disobedience. Examining this trend comprises the focus of Chapter 4. As I show, many climate disobedients feel compelled by the lack of effective political action to take matters into their own hands, and act so as to directly prevent climate harm themselves—above all, the very great harms of catastrophe and collapse. Illustrative here is Extinction Rebellion's "Declaration of Rebellion":

> When Government and the law *fail to provide any assurance of adequate protection*, as well as security for its people's well-being and the nation's future, it becomes the right of its citizens to seek redress in order to restore dutiful democracy and to secure the solutions needed to *avert catastrophe* and protect the future. *We hereby declare the bonds of the social contract to be null and void*, which the government has rendered invalid by its continuing failure to act appropriately. We call upon every principled and peaceful citizen to rise with us.[83]

[83] Emphasis added; Extinction Rebellion 2022.

In other words, for these activists, disobedience is understood as a justified and even necessary response to the continuous governmental failure to take action against the credible threat of climate catastrophe. Such a dereliction *merits* disobedience insofar as, again, any authority that fails to ensure basic safety and security lacks legitimacy, and thus too a valid claim-right to obeisance; and it *necessitates* disobedience in the sense that, if governments are unwilling to act, then it falls to the people themselves to serve as the final guarantors of safety, even if this means operating outside, or in direct violation, of the law. Preventing catastrophe is simply too important: particularly in the climate case, the risk too great, the stakes too high. As another disobedient group, Just Stop Oil, puts it: "If we continue down our current path it will destroy families and communities. We will face the starvation and the slaughter of billions of the poor—and the utter betrayal of our children and their future."[84] Many climate disobedients feel it is up to them to stop this from happening.

Understanding civil disobedience in this way—as aiming directly and in the first instance at the prevention of harm—cuts against a virtually universal consensus in the literature, which holds that civil disobedience is, invariably, a *communicative act*. In other words, the consensus view is that the essential ends of civil disobedience are to expose injustice, express moral condemnation, and persuade the broader public to reform. Chapter 4 rejects this rigid identification of civil methods with communicative ends. For, while some cases of climate disobedience certainly have communicative ends, many others do not; rather, they aim narrowly, or at least primarily, at the prevention of harm.

The distinction between "harm-preventative civil disobedience" (HP-CD) and communicative civil disobedience (C-CD) has application beyond the climate movement. Thus Chapter 4 develops these concepts as part of a more expansive theory of civil disobedience. Central here is the claim that, while HP-CD and C-CD share many features, one critical difference between them is how each instrumentalizes the penal process. This is a subtle but significant point. Consider that, according to many civil disobedience theories, voluntarily accepting punishment offers a strategically valuable means of furthering the communicative ends of the protest, for instance by eliciting sympathy for the disobedients (and thus their cause). This logic breaks down, however, when at the end of disobedience is the prevention of harm, rather than communication: in such cases, there is no clear instrumental reason to accept punishment. In this sense, it is perhaps unsurprising that many climate disobedients have resisted punishment by pleading not guilty on the grounds that they acted out of "necessity."[85] Paradoxically,

[84] Just Stop Oil 2022b. Many other climate disobedients and activist organizations have expressed similar views. See Chapter 4, esp. sections 4.3 and 4.6.

[85] Broadly, the necessity defense is a formal legal excuse applied in cases in which an illegal action was taken to prevent exigent and more severe harms. I further explain this defense, and its role in HP-CD cases, in Chapter 4.

however, many of those same disobedients reported themselves to the authorities and awaited arrest, and some have expressed "disappointment" at prosecutors dropping charges against them.[86]

This paradox dissolves, however, when we understand that while disobedients in HP-CD cases are not instrumentalizing *punishment*, they are nevertheless instrumentalizing the *penal process*, precisely by pleading necessity. For successful use of the "necessity defense" establishes a legal precedent for regarding their and similar acts of disobedience as excusable responses to the threat of climate catastrophe. This, in turn, affords future disobedients greater assurance that their action will not be met with jail time, and thus provides a means by which such acts may proliferate. The ultimate aim, then, in using the necessity defense is to reclaim a form of public authority over the agents and practices that threaten the well-being of the polity. Thus, much as "necessity" serves a justificatory basis for governments to employ authoritarian emergency powers in confronting the climate crisis, so too can "necessity" ground ordinary citizens' use of extralegal actions when adequate governmental action is lacking.

Some may question whether civil disobedience—or individual action of any sort—is actually capable of preventing climate harm, and thus whether invocations of "necessity" are plausible in this context. However noble individual acts of disobedience may be, a critic might argue, it seems highly implausible, if not altogether incoherent, to believe that they could meaningfully prevent climate harm—and least of all the very great harm of politically catastrophic climate change. In this sense, regarding climate disobedience as a genuinely *harm-preventative* act would seem to be misguided. Much of Chapter 4 is devoted to responding to this objection. It does this by drawing on a surprisingly analogous debate over the significance of individual votes in mass elections, which have also been described as no more significant than single drops of water "in a vast sea."[87] The argument there is complex, but, broadly speaking, I claim that (i) so long as a given act of civil disobedience serves to bring us closer to meeting emissions-reductions goals; and (ii) so long as we ultimately achieve those goals; then (iii) there is a coherent basis for regarding the disobedient act as having been causally necessary for averting catastrophe. Thus, the use of the necessity defense in such cases is appropriate.

Overall, the core chapters of this book (Chapters 2 to 4) attempt to elucidate just how much climate change threatens prevailing political values, practices, and institutions, and thus just how urgent attending to the crisis is, *politically speaking*. Although presented in various forms, the essential claim throughout is that preventing politically catastrophic climate change is of such overwhelming importance that it warrants departures from ordinary moral and legal constraints:

[86] Associated Press 2018; Hand 2018.
[87] See Chapter 4, section 4.4.1.

ultimately, we must take action, even if our only options for doing so are in some way unfair, undemocratic, illiberal, or illegal.

This is a difficult view to defend. But we can no longer have any doubt: failure to take adequate action at this late stage will entail our eviction from the veritable Eden of our current geological epoch—the Holocene, exceptional for its enduring climatic stability and natural abundance. If we are to avoid descending into a world marked by conflagrations, conflict, destruction, disease, and death—a world in which we can have no guarantee of civilization or even survival—we have no choice but to act "as if our lives depended on it," as Greta Thunberg has so compellingly pleaded, and with all the political force and urgency we can command.

2
Climate Change, Catastrophe, and the Circumstances of Justice

> If we continue on our current path we will face the collapse of everything that gives us our security: food production, access to fresh water, habitable ambient temperature, and ocean food chains. And if the natural world can no longer support the most basic of our needs then much of the rest of civilization will quickly break down. Please make no mistake: climate change is the biggest threat to security that modern humans have ever faced.
>
> Sir David Attenborough, 2021 UN Security Council speech[1]

2.1 Introduction

It is a tragic feature of human cooperation that the most efficacious allocation of burdens is not always the fairest. In contemporary political thinking, fairness holds so much weight that we often believe it better to abandon a common project than to accept an inequitable distribution of benefits and burdens. Yet, in some cases, the ends of our shared endeavors are not elective, or are so valuable that we may think cooperation should proceed regardless of fairness concerns. This appears true of warfare and diplomacy, for instance, and often also for healthcare. In such cases, we may *hope* that our shared goals can be met without unduly or arbitrarily burdening any particular party. Nevertheless, trade-offs are often unavoidable, such that we must sacrifice either some amount of fairness if our goals are to be met, or of efficacy if uncoerced cooperation is to continue. A fundamental political problem, then, is determining which should take priority in a given situation.

In this chapter, I argue that the imperative of preventing climate catastrophe represents another example of this dilemma. After decades of inaction, we may no longer be able to limit some of climate change's most harrowing effects—including sea-level rise, extreme weather, the proliferation of mosquito-borne diseases, mass extinctions, agricultural collapse, and water scarcity—without imposing unfair

[1] https://youtu.be/MaweqwsN62k

Climate Change as Political Catastrophe. Ross Mittiga, Oxford University Press.
 DOI: 10.1093/oso/9780192868879.003.0002

burdens on some parties. This is particularly clear in the intergenerational context: there is likely no way to prevent climate catastrophe unless those of us alive today make substantial sacrifices for the sake of future people, who are widely expected to be wealthier than we are now. But the same may also hold *intra*generationally: effective precautionary action against climate catastrophe may require imposing significant costs or developmental delays on relatively poorer states like China and India, for which they may never receive full compensation.

The tension between fairness and efficacy in the context of climate change has not been very well explored in the literature.[2] It therefore remains unclear under what conditions, or to what extent, prioritizing effective precautionary action over distributive fairness could be justifiable. This comprises the focus of the present chapter.

I begin, in section 2.2, by considering the possibility that *no* trade-offs between fairness and efficacy are actually necessary. This view is popular in the intergenerational-justice literature; it is staked on the idea that current generations should simply defer the costs of presently undertaken action onto richer, future people in the form of public debt. For various reasons, I argue that this proposal fails.

In section 2.3, I turn to an alternative approach developed by Simon Caney. This approach acknowledges the potential for conflict between the goals of efficacy in avoiding harm and fairness in distributing burdens, and offers a sophisticated account of when we might justifiably prioritize the former over the latter. What results is an improvement over the intergenerational-justice approach. I argue, however, that Caney's proposal also falls short in several key respects.

I then advance a new position in sections 2.4 and 2.5. Broadly, I argue that prioritizing efficacy over fairness is justified when societies are faced with the prospect of "political catastrophe." Political catastrophe occurs whenever material scarcity is so extreme that fair social cooperation and political stability become impossible to realize or sustain. My essential claim is that preventing the loss of the conditions that make justice possible ought to be given greater weight than satisfying any particular instance or application of justice. Defining catastrophe in relation to the circumstances of justice helps explain how precaution coheres with broader theories of justice, making my argument less ad hoc than alternatives. It also clarifies the specifically *political* stakes of precaution: we should take action both to avoid harm *and* to prevent social-institutional collapse. The prospect of collapse provides a strong moral and prudential base of motivation for states and other temporally unbounded actors to act immediately and decisively against credible threats.

[2] There are some exceptions—perhaps most notable is the work of John Broome (especially Broome 2012). See also Posner and Weisbach 2010 and Budolfson 2021. I discuss Broome's view at greater length later in the chapter.

In section 2.6, I argue that climate change threatens political catastrophe and, given this, those of us alive today are obligated to take all reasonable measures to address it—including reducing emissions, enhancing carbon sinks, supporting adaptation measures, and ensuring that losses and damages do not precipitate extreme scarcity. Moreover, I claim that this obligation must be fulfilled even if doing so conflicts with ordinary principles or rules of justice.

2.2 Intergenerational-justice approaches

One well-discussed example of conflict between fairness and efficacy in the context of climate change appears in the intergenerational-justice literature. This stems from the fact that virtually every economist expects future generations to be wealthier and more technologically advanced than we are now.[3] If future generations *will* be better off, then imposing strenuous climate-action burdens on current generations seems unfair, as it would amount to requiring poorer people to sacrifice for the benefit of richer ones.[4]

We can call this the "intergenerational equity objection" to immediate climate action, of which there are many examples in the literature. Robert Lind, for instance, asks: "Can we justify current generations sacrificing 2–3% of GWP to increase the wealth of future generations who even after deduction for the high damage scenario are 2–15 times richer than the present generation?"[5] Others have argued similarly that, even accounting for the likely costs of climate change, "it will be far more expensive to cut carbon-dioxide emissions radically than to pay the costs of adaptation"[6]; accordingly, justice would be better served by devoting our scarce resources to assisting the global poor alive today—even if this means continuing fossil fuel-intensive development.[7]

[3] Schelling 2000; Dasgupta 2008: 6; Godard 2008: 33. For discussion, see Rendall 2011: 884ff.; cf. Posner and Weisbach 2010: 77–9.

[4] Of course, not everyone takes this view. Broome (2012: 45ff.), for instance, argues that imposing burdens on current generations would be justified in accordance with the polluter-pays principle (PPP), because they are contributing to the problem while future generations are not. There are a couple of problems with this view, however. First, if applying the PPP compromises minimal standards of living, as it might do in (e.g.) China, India, and other high-emitting developing states today, then an (unqualified) intergenerational PPP would likely generate as much unfairness as it attempts to resolve. Moreover, while the PPP may be able to establish that current generations should pay (at least some of) the costs associated with reducing emissions, it cannot as easily establish duties for another crucial aspect of mitigation: maintaining and enhancing carbon sinks, like rainforests (Armstrong 2016; Page 2016; cf. Honegger, et al. 2021). Thus, because an intergenerational PPP risks unfairness in an intergenerational context (for its failure to take economic considerations into account), and because it is incomplete with respect to climate duties, I set it aside here. This of course is not to deny that a defensible, *intra*generational application of the principle could be developed (as I argue in Mittiga 2019a).

[5] Lind 1995: 384; quoted in Rendall 2011: 886.

[6] There are strong reasons to doubt this claim, as I explain below.

[7] Lomborg 2001: 318, 310–18. For additional statements of essentially the same objection, see Nordhaus 1997, 2008a, 2008b; Lomborg 2007. But cf. Cole 2003; Caney 2009a, 2010; and Gardiner 2011: 248 and passim.

Political theorists and climate ethicists have developed various responses to the intergenerational equity objection, all of which aim at making space for effective precautionary action without undercutting the priority of fairness.[8] Most commonly, scholars argue that even if we grant that future generations will in fact be richer,[9] imposing climate action burdens on those alive today would nonetheless be justified *so long as* the costs of present action can be shifted onto the future (e.g. through heritable debt).[10] In other words, by separating the costs of climate action from the actions themselves, we can ensure that the (richer) future pays, and thereby preserve intergenerational equity. Any apparent conflict between effective precaution and intergenerational fairness would thus disappear.

While on first blush this proposal seems plausible, on closer inspection a number of issues arise. Most pressingly, significant doubt exists about whether shifting costs onto future generations is actually possible. Many economists argue that wealth can only be transferred between contemporaries, not generations.[11] As Matthew Rendall explains: "it does not seem possible for the present generation *taken as a whole* to borrow money from the future," because any "loan made now comes out of today's spending and investment, and will in turn be repaid to future people."[12] If this is right, then any attempt to borrow must clarify which contemporary agents should be the debtors and which the lenders, and on what terms. Doing this while preserving intergenerational equity would require disaggregating generations, sorting agents with differing debt-bearing capacities into distinct economic strata, producing some reliable estimate of how those agents' capacities will change over time (e.g. will they continue occupying the same economic strata?), and, finally, selecting a global-justice principle that takes these factors into account when allocating burdens.[13]

[8] Another line of response challenges the damage-estimate models that economists use. I am not qualified to assess this approach, suffice to say that elements of these models certainly do appear questionable: e.g. the assumption that damage costs of low-level warming (like 0.5°C) can be scaled up linearly (rather than exponentially) to provide estimates for greater increases in temperature (Wagner and Weitzman 2016: esp. 60ff.). Similarly, the common tendency to exclude "fat-tailed uncertainties" from damage estimates seems defensible only as a simplifying assumption (Rendall 2011: 889; Furman, et al. 2014; IPCC 2014b: 246; Wagner and Weitzman 2016; see also McKinnon 2012: 5), and may serve to introduce a (potentially radical) conservative bias (Oreskes, et al. 2019).

[9] Some theorists doubt this assumption, pointing out that growth in the costs associated with climate change may well outpace general economic growth over the relevant timescales. I address this below.

[10] Some variant of this proposal has been advanced by a number of ethicists, including Matthew Rendall (2011), Simon Caney (2012, 2014a), John Broome (2012), and others. Various economists have at some point advocated the same, including William Nordhaus (2008a: 179–80) and Nicolas Stern (2007). For a helpful, interdisciplinary defense of (a version of) this position, see Kelleher 2015.

[11] See e.g. Ferguson 1964; Cooper 1986; Labonte and Makinen 2005, all cited in Matthew Rendall's (2011) helpful discussion.

[12] Rendall 2011: 892; citing Buchanan 2004: 324–5. See also Broome 2016.

[13] This problem holds even if heritable debt is not the mechanism of intergenerational wealth transfer. For instance, Broome and Foley (2016) have proposed diverting future-oriented research and development investments into an international climate bank that could then use the funds to finance mitigation policies. But as this would similarly require some countries to forego welfare gains for the benefit of others, the same question arises: who should sacrifice now and why?

Suppose for the sake of argument that this can be done or that some other mechanism for fairly deferring the costs of present action onto future generations can be found. Two problems remain. First, even if we could defer *some* of the costs of present climate action onto the future, it is nonetheless impossible to defer *all* of the costs.[14] For instance, effectively addressing climate change will ultimately entail many workers losing their jobs—particularly in the fossil-fuel and animal-agriculture industries. Yet one's occupation is often an integral part of one's identity and sense of self-respect. (This may be especially true for those in iconified industries, like farmers or Appalachian coal miners.) Moreover, many displaced workers may have limited ability to adopt new professions. These costs and sacrifices cannot be deferred, and pecuniary compensation can only go so far. Thus, some residual injustice remains. While this cost of course pales next to that of climate catastrophe, so long as it cannot be fairly distributed, it represents another example of the tension between fairness and efficacy.

We might also worry that any plan that proposes transferring costs onto the future ultimately depends on unverifiable assumptions about future economic conditions. Should it turn out that future generations are poorer, or that growth in the costs associated with climate change outpace general economic growth, then deferring even some of the costs will turn out to have been unfair. When we defer costs, we take a gamble with justice that might not bear out.

This suggests a broader problem: to justify any particular program of action, intergenerational-justice approaches must regard certain assumptions about the future as matters of fact. This is necessary for specifying unambiguously what each generation owes to future ones. For our purposes this means that, to require current generations to act, intergenerational-justice theorists must maintain that the future effects of unmitigated climate change will, *in fact*, be grave. Regardless of scientific consensus about the anthropogenic origins of climate change, however, it remains impossible to predict precisely what (or when) climatic changes will occur, or how such changes will impact human well-being. Consider, for instance, the IPCC's prediction that a doubling of atmospheric CO_2e[15] will "likely" increase global mean temperature by 1.5 to 4.5°C.[16] If warming falls at the low end, many (though not all) states should be able to adapt without suffering catastrophic losses. Warming at the high end, however, is likely to prove devastating for ecological, economic, and political systems throughout the world. This variability weakens intergenerational-justice theorists' ability to insist on a particular allocation of

[14] This point, along with the following example, draws on Caney 2014b: 133–4.

[15] That is, "carbon dioxide equivalent," a shorthand metric representing all greenhouse gases.

[16] Scientists use the term "climate sensitivity" to describe the total global mean temperature increase after a doubling of CO_2e. Preindustrial CO_2e was 280 ppm; at the time of writing (October 2022), it is over 412 ppm. Without additional emissions reductions, it may exceed 700 ppm by 2100. IPCC 2013: 16; IPCC 2014b: 13. See also Wagner and Weitzman 2016: 53.

climate-action burdens.[17] For if we were to *know* that temperature increases will be low, present generations could justifiably leave the bulk of climate action to the (likely wealthier and more technologically advanced) future. Conversely, if we were certain that temperature increases will be great, the case for imposing stringent burdens on present generations would be much stronger.[18]

A related issue is that, in the absence of certainty, the normative motivation for immediate climate action on intergenerational-justice approaches goes unstated. Most likely, intergenerational-justice scholars wish to reduce the likelihood that climate change will overwhelm the response capacities of future generations, leaving them with an uninhabitable planet. But if this is right, then these scholars are implicitly relying on a notion of precaution without considering the implications this might have for their other theoretical commitments—including the potential tensions that realizing effective measures may have with standards of fairness.

Even if solutions to these various problems could be devised, we might still worry that the possibility of deferring costs provides only a contingent (and thus precarious) justification of contemporary precautionary climate action. For, on this view, immediate action is permissible *if and only if* present generations can in fact shift costs onto the future. Such a qualified position is likely to be unsatisfying for anyone centrally concerned with preventing catastrophe.

2.3 Prioritizing precaution: Caney's approach

Recent work by Simon Caney offers another way of navigating conflicts between fairness and efficacy in responding to the climate crisis. Caney starts with "the assumption that it is of paramount importance that humanity avoids dangerous climate change," and acknowledges that it may be impossible to achieve this without upsetting distributive justice.[19] He then attempts to specify the conditions under which effective preventative action justifiably takes priority over fairness.[20]

Central to Caney's argument is a distinction between "first-" and "second-order responsibilities." First-order responsibilities cover core climate duties: to "mitigate climate change," "enable adaptation," "compensate people for harm done," and to pick up others' slack in instances of non-compliance.[21] While the claim that such duties exist is uncontroversial, Caney's account leaves us with a number

[17] Some have contested this, arguing for a more stringent approach to future discounting (which would be central to determining the appropriate allocation of present burdens); see e.g. Fleurbaey, et al. 2018.

[18] Notably, the timing and form of climate changes are just two areas of uncertainty. Others concern the physical effects of climate change and their costs.

[19] Caney 2014b: 127–8.

[20] Caney (2014b) presents this as a trade-off between two kinds of climate justice: one focused on ensuring that burdens are fairly distributed, the other focused on the avoidance of serious harms.

[21] Caney 2014b: 134, 135.

of unanswered questions. For one, it does not specify to which agents first-order responsibilities fall, although, considering Caney's earlier work, it is likely this would be some construal of "the most advantaged"[22]—in particular, the most advantaged states.[23] Caney also does not indicate clearly what the normative foundations for these responsibilities are, though, again judging by his earlier work, the imperative to prevent harm or the erosion of just entitlements (to health, safety, and a decent standard of living) likely fulfills this role.[24] Finally, and most pressingly, Caney's account does not specify under which conditions, if any, first-order responsibilities to prevent climate catastrophe should take priority over ensuring that burdens are distributed fairly.[25]

These ambiguities owe to the fact that Caney is principally concerned with "second-order responsibilities," which are those agents have (a) to ensure that others (presumably states) comply with their first-order responsibilities, and (b) to promote a normative-institutional context in which preventing climate catastrophe is likely to succeed.[26] Caney's priority argument is meant to apply only to these second-order responsibilities. His essential claim is that those who have the *power* to prevent dangerous climate change (in the second-order sense) have a *responsibility* to do so. He refers to this as the "power/responsibility principle" (hereafter PRP).[27] The PRP applies when the following, jointly sufficient conditions obtain:

(1) "humanity faces a prospect of disastrous harms";[28]
(2) certain agents have the capacity to "reduce, or severely limit, the chances of these dire outcomes" coming to pass;[29]
(3) moreover, these agents are *uniquely* able to prevent catastrophe—which means both that others lack similar capacities and that "if disaster is to be averted, [all qualified agents] must act";[30]
(4) there are no "sufficiently weighty countervailing considerations" that "take priority" over the responsibility to act.[31]

Caney argues compellingly that the first three conditions are satisfied with respect to climate change, and offers a plausible defense of (4). But a number of issues remain.

[22] See Caney 2005: 769. This reference is suggested by Caney 2014b: 135n.32.
[23] Caney 2005: 751.
[24] This is suggested by his focus on avoiding harm, and (in earlier work) the impact of climate change on core human rights (Caney 2009b).
[25] Caney may believe that a trade-off between fairness and efficacy in the context of first-order responsibilities is unnecessary.
[26] He formulates (b) quite capaciously to include "norm entrepreneurship," disobedience, incentivization, enablement, and undermining resistance to action (Caney 2014b: 136–8).
[27] Caney 2014b: 141.
[28] Caney 2014b: 142.
[29] Caney 2014b: 143.
[30] Caney 2014b: 143.
[31] Caney 2014b: 143–4.

First, and most critically, we are looking for a priority argument that applies to core—i.e. first-order—climate duties. Caney repeatedly insists, however, that his argument applies *only* to second-order responsibilities.[32] His reasoning for this appears to be that first-order responsibilities can be distributed in accordance with a principle of corrective justice (for past emissions), and so no trade-off between fairness and efficacy is necessary.[33] Yet so long as it is at least possible that the most efficacious distribution of climate-action burdens will not be the fairest or must just (however construed), our original problem remains. Whichever justice principle we employ, we must still ask: under what conditions and to what extent should efficacy of action take priority over fairness in the context of preventing climate catastrophe?

One way to answer would be simply to extend Caney's priority argument to cover first-order responsibilities. We might stipulate that, whenever the four conditions noted above obtain, all states (or other relevant actors) with the power to prevent catastrophic climate change (in a first-order sense) also have the responsibility to do so. Granting that this extended PRP is possible, we can consider what problems, if any, it faces.[34]

Starting with condition (1), Caney is right to argue that climate change portends serious future harms if we fail to take decisive and immediate action. But whenever we speak about the duty to prevent *future* harms, some account of how probability affects responsibility is needed. At what point, exactly, does the threat of climate catastrophe generate a duty to act? Does the strength of one's responsibility scale up with increased likelihood? Or is it categorical—i.e. does one have an obligation above a certain minimal threshold of probability? Caney does not address these issues.

Regarding conditions (2) and (3), two problems arise. First, particularly with respect to first-order responsibilities, every capable agent is *not* needed for effective climate action; a critical mass would suffice. This is especially clear if we consider that in 2018 the top ten state polluters alone were responsible for more than 68 percent of global emissions, while the bottom hundred countries accounted for only about 3.6 percent.[35] Of course, many of the countries between the top ten and bottom hundred have the power to reduce their emissions substantially. But their reductions are far less important for preventing catastrophe; they may even be unnecessary if the top polluters act.

Caney later concedes that, even for second-order responsibilities, "it may be too strong to say that the intervention of all [capable] agents is necessary," but he

[32] Caney 2014b: 135, 142, and *passim*.

[33] Caney 2014b: 140–1.

[34] It is worth emphasizing that the following is not a critique of Caney's account, per se, as he does not attempt to extend the PRP as I am proposing. That said, some of the problems noted may also apply to his priority argument for second-order responsibilities.

[35] The US, China, and the E.U. alone produced about 42 percent of global emissions that year (Friedrich, et al. 2020).

holds nonetheless that "[w]hether dangerous climate change is averted ... depends on whether a sufficient number of ... agents take up these roles."[36] The problem with this is not only the ambiguity surrounding what counts as a sufficient number; Caney's admission also undercuts the basic motivational force of the PRP, which holds that "acting on it is *necessary* to protect those whose interests are threatened."[37] Even accepting the (sound) assumption that the actions of some capable agents *are* necessary to prevent catastrophe, in most cases, complying with the PRP will *not* be necessary for any particular agent. Thus, we have a collective action problem. If the concerted action of a certain number of capable agents is necessary to avert disaster, but the participation of any particular agent is unnecessary, then individuals have considerable incentive to free-ride.[38] Without a coordination mechanism or a principle that can allocate responsibility in instances of redundancy, how can the PRP be effective?[39]

This problem becomes more worrying if we accept that fulfilling climate duties may come at a cost to an agent's other important values or ends. Then, instead of just lacking an incentive to comply, agents will have an incentive *not* to comply. Perhaps in recognition of this, condition (4) stipulates that agents possess no "sufficiently weighty countervailing considerations" that "take priority" over their responsibility to act. While this *may* hold true for second-order responsibilities, it almost certainly does not for first-order ones. Reducing emissions, enhancing sinks, establishing and financing compensation mechanisms, enabling adaptation, and so on, are difficult and expensive tasks. This is especially true for poor states, for which general climate inaction and fossil fuel industrialization are often still regarded as expedients to essential developmental gains.[40] More importantly, even if costs were not a concern, we still might worry that condition (4) is tautological and ambiguous: tautological in the sense that it stakes the priority of preventative action on the ground that there are no considerations that take priority over preventative action (and thus asserts what it is meant to prove); and ambiguous in the sense that it does not indicate clearly what kind of consideration, if any, would be sufficiently weighty so as to overpower an agent's duty to act.

[36] Caney 2014b: 143.

[37] Emphasis added; Caney 2014b: 144.

[38] This holds even at the level of second-order responsibilities: if agents X, Y, and Z all have the power to enforce compliance with first-order responsibilities (say, to reduce emissions), but the actions of one or two agents is sufficient to accomplish this, then X, Y, and Z all have an incentive to shirk their duties and hope that the others comply.

[39] To see the problem this poses for motivation, consider a modified version of Peter Singer's (1972: 231ff.) famous drowning baby example, which—as Caney notes—relies on a structurally similar line of reasoning (Caney 2014b: 146n.47). If I was not alone in walking by the pond in which the baby was drowning, but was accompanied by many people capable of rescuing it, who has the final responsibility to act, and why? If we all have this duty, but only one person is required to act to save the baby, how should we decide who should wade in?

[40] Cass Sunstein (2006: 882) argues this point forcefully against Stephen Gardiner.

Thus, while Caney's account is laudable for directly addressing the essential problem—how should we balance efficacy and fairness in the context of preventing dangerous climate change?—it suffers from a number of critical issues when extended to cover core precautionary duties, like reducing emissions. Consequently, we are still in need of an account that explains under which conditions, and to what extent, the value of efficacy may justifiably take priority over fairness when attempting to prevent climate catastrophe.

2.4 Alternative justificatory bases for prioritizing precaution

Over this section and the next, I present a new approach. I argue that prioritizing efficacy over fairness is justifiable whenever the conditions that make justice and political stability possible are themselves compromised or credibly imperiled. I refer to such situations as political catastrophes, a concept introduced in the introductory chapter and further elaborated later in this chapter.

2.4.1 Catastrophe

The idea of "climate catastrophe" is frequently invoked when attempting to motivate the need for (or otherwise justify) precautionary action. But what catastrophe signifies is not always clear. The most recent OED entry defines it as a "sudden disaster, wide-spread, very fatal, or signal." Cass Sunstein, on the other hand, describes catastrophes as instances of harm that "involve a large number of human deaths."[41] Perhaps more stringently, Richard A. Posner defines them as events that "threaten the survival of the human race."[42]

The focus on death in these and many other accounts[43] accentuates the moral relevance of precautionary action. Yet we might wonder if death is a sufficient or even necessary feature of catastrophe. Are not instances of extreme suffering catastrophic, before or apart from human death? What about cases of severe political instability, in which norms, cooperation, and institutions break down?

What I call *political catastrophe* captures these omitted elements.[44] Political catastrophe is not defined by a body count, though it may follow from, or result in,

[41] Sunstein 2006: 869.

[42] Posner 2004: 6.

[43] One notable exception is the economics literature, which typically defines catastrophe in terms of sudden and sharp drops in consumption or GDP (at times referring to them simply as "consumption disasters" [e.g. Barro 2009]); even in that literature, though, some influential scholars are advocating the foregrounding of mortality in conceptualizing and modeling catastrophe (e.g. Martin and Pindyck 2021).

[44] Even if these phenomena fall outside of catastrophe, as conventionally understood, we might still think them sufficiently grave as to warrant precautionary action.

human death. Rather, political catastrophe is characterized foremost by a state of extreme material scarcity, in which one can only meet one's basic needs by denying another (or others) the ability to do the same. To count as political catastrophe, scarcity must be enduring, though not necessarily permanent or irreversible; it must be irremediable by simple local efficiency gains or redistributions; and it must afflict at least a sizable minority of a given population.[45] When extreme and enduring scarcity of this sort occurs, it becomes impossible to realize or sustain relations of justice and, by extension, uncoerced forms of social cooperation, all of which critically endangers the stability of democratic governance structures (or any governance structure at all).

Central to the idea of political catastrophe, then, is something like John Rawls's "objective circumstances of justice," which he defines as "the conditions under which human cooperation [is] both possible and necessary."[46] The most important of these circumstances is "the condition of moderate scarcity," which obtains when "[n]atural and other resources are not so abundant that schemes of cooperation become superfluous, nor are conditions so harsh that fruitful ventures must inevitably break down."[47] In describing the latter condition, that of extreme scarcity, Rawls refers to a passage by David Hume, which describes a society that has fallen

> into such want of all common necessaries, that the utmost frugality and industry cannot preserve the greater number from perishing, and the whole from extreme misery; it will readily ... be admitted, that the strict laws of justice are suspended in such a pressing emergency, and give place to the stronger motives of necessity and self-preservation.[48]

This passage makes vivid the idea that, without adequate material resources, justice becomes difficult (if not impossible) to realize or sustain. Without redress, such extreme scarcity inevitably leads to social and political collapse.[49]

I argue that the imperative of responding to political catastrophe sanctions deviations from ordinary moral constraints—not the complete and self-centered abandonment of justice that Hume portends,[50] but rather shifts in the allocation of burdens and benefits necessary to alleviate or resolve the catastrophe quickly and effectively. I develop a justification for this claim and address some potential

[45] One might wonder what counts as "local," and who comprises the relevant "population." For present purposes, we may assume that local pertains to any well-defined, geographically continuous political-administrative unit (like a state), and the relevant population to all human beings residing in that political unit at a specific point in time.

[46] Rawls 1999: 109.

[47] Rawls 1999: 109, 109n.3, 110.

[48] Hume 1751: sec. III, pt. I.

[49] The link between food and/or water scarcity and political collapse is helpfully expounded in Tainter 1989. For further discussion, see Chapter 1 and the Appendix.

[50] Except, perhaps, in the most extreme political catastrophes. On this, I follow Barry 1978.

objections to it in sections 2.4.2 and 2.6, respectively. But we should note here that something like this idea is already widely, if often only tacitly, accepted in various contexts. Medical triage offers a clear example.[51] Under normal conditions, clinicians are expected to "devot[e] maximum time and resources to the sickest patients,"[52] without respect to morally arbitrary factors like age, weight, expected recovery time, etc. Yet, when medical supplies are limited and/or services are under extreme duress, these expectations change, and factors previously considered arbitrary for allocating medical resources may become determinative.[53] For instance, in the case of vaccine shortages, officials may justifiably discriminate based on age, prioritizing the very young and the very old. Likewise, in military triage situations those patients most likely to recover quickly and resume action may justifiably be prioritized.[54] In short, under conditions of extreme scarcity, the principles, rules, and values that normally govern the dispensation of medical care are justifiably outweighed by—and are subordinated to—higher-order concerns like saving the most lives, protecting the most vulnerable, or maintaining the war effort.

I claim that the same dynamic holds true in more general instances of extreme scarcity: the principles and values that normally guide the moral-political calculus—especially distributive fairness—may justifiably be subordinated *if* and *to the extent that* doing so is necessary to restore conditions of moderate scarcity, which constitute the material basis of justice. In this sense, political catastrophe may be used to define genuine states of exception because it gives content to the concept of "necessity":[55] making exceptions to moral principles or the legal order is legitimate when doing so is necessary to preserve or restore the conditions that sustain them.[56]

2.4.2 Two priority arguments

Accepting all of this, however, we must still ask: are deviations from moral principles justified to prevent *future* political catastrophes? (After all, mitigating *risk* is at the core of *pre*caution.) And if such deviations are so justified, what makes them so and to what extent?

[51] This argument is developed at length in Adams and Mittiga 2021. For an important earlier treatment with application to climate change, see also McKinnon 2012.

[52] Moskop and Iserson 2007: 282.

[53] For an engaging and accessible discussion of this dynamic at a broader, political level, see Mulgan 2014.

[54] Moskop and Iserson 2007: 283.

[55] This claim picks up on perceptive remarks in Lazar 2009, esp. ch. 4: e.g. "whatever means we come to for understanding the ethics of emergency ... we must give moral content to the idea of necessity"; and "Emergency action, when it is justified at all, is justifiable on the basis of a further value: the preservation of the state order" (Lazar 2009: 88).

[56] The following chapter provides an extensive defense of the claim that politically catastrophic climate change may legitimate states of exception and the use of authoritarian emergency powers.

2.4.2.1 Contractualist argument

One potential means for justifying the prioritization of efficacy over fairness in addressing risks of political catastrophe is by adopting an impartial decision-making framework that melds the perspectives of justice's beneficiaries and burden-bearers.[57] We can model impartiality within such a framework by prescinding information about the place or time to which each person taking part in the decision-making process belongs.[58] Therein, participants should assume that the actions or inaction of some agents can precipitate extreme scarcity for others, and thus that political catastrophe must be deliberately avoided.[59] Ignorant of their temporal or spatial locations, rational, self-interested agents occupying such a framework would, I argue, seek to ensure that every generation takes all reasonable measures to reduce the likelihood of political catastrophe.

Of course, preventative action is not without costs.[60] Thus these agents will also want to ensure that the burdens on any person or group are not so great that they themselves artificially induce political catastrophe. Deliberators could therefore be expected to reject an austerity program that forces extreme scarcity in the present simply to avoid the same later on.[61] Yet it would be rational for them to accept any necessary burdens falling below this upper limit. If this is right, then from an impartial perspective—one that represents the interests of all people and generations[62]—imposing strenuous (but not politically catastrophic) burdens on some to prevent political catastrophe from befalling others is morally justified. Put another way, from the perspective of the least-advantaged members of the least-advantaged generation (i.e. those facing political catastrophe),[63] precautionary efficacy justifiably outweighs fairness. Thus, should it be the case that protecting future generations from political catastrophe is possible only by imposing stringent burdens on earlier, poorer generations, doing this would be justified.

Future generations consigned to previously preventable political catastrophe would likely find any other conclusion unacceptable. For them, it would be no consolation at all to be told that their plight is simply what fairness required. Anyone at risk of occupying such a position would undoubtedly prefer an *unfair* schedule of burdens[64] that prevents political catastrophe to a *fair* one that does not.

[57] We should consider both when structuring intergenerational relations, as both Shue (1996: 164–6) and Caney (2005: 762) stress.

[58] Much like Rawls's original position.

[59] While this seems to modify the original position, Rawls may tacitly stipulate similar constraints. Consider where he refers to a "basic needs principle" in later work (Rawls 2005: 7, 166, 228–9). For an attempt to extend this principle to climate change, see Wolf 2009.

[60] Sunstein 2006: 879, 881, and *passim*.

[61] Such a program violates Rawls's "strains of commitment" argument, which stipulates that deliberating parties should reject agreements they "can adhere to only with great difficulty" or that "may have consequences they cannot accept" (Rawls 1999: 153).

[62] Rawls 1999: 258.

[63] Gardiner (2006) argues that climate catastrophe is a worst-off outcome, in the Rawlsian sense, and should be avoided through precaution. See also Wolf 2009; Rendall 2011; McKinnon 2012: 32ff.

[64] Say one that assigns heavier burdens to earlier and/or poorer generations *simply because* they are earlier or poorer.

2.4.2.2 Paradox of justice

Another way of justifying the priority of precautionary efficacy over fairness is suggested by Karl Popper's famous "paradox of tolerance," which holds that "[u]nlimited tolerance must lead to the disappearance of tolerance." The essential idea here is that if a society tolerates the expression of all beliefs, including intolerant ones, it may be unable to sustain tolerance in the long run—particularly if the intolerant views proliferate. In such a case, "the tolerant will be destroyed, and tolerance with them."[65] Rawls shares Popper's view, arguing that if "the tolerant [in a given society] sincerely and with reason believe that their own security and that of the institutions of liberty are in danger," they may justifiability "restrict the freedom" of "intolerant sects."[66]

A similar paradoxical relationship appears to hold with respect to justice. That is, if the only way to realize or act upon a particular conception of justice entails causing or allowing political catastrophe (which, again, constitutes a dissolution of the conditions that make justice possible), then it follows that sometimes remaining just may only be possible at the cost of justice in general. It therefore seems reasonable to assume that we may limit or subordinate particular applications of justice when doing so is necessary to ensure the continued general viability of justice (and, with it, political stability). Below, I argue that the climate crisis represents just such a case: that is, that the demands of intergenerational or international justice may undermine our ability to prevent political catastrophe and thus sustain the conditions that make justice possible. Consequently, justice may require accepting some amount of injustice.

2.5 What does effective precaution entail?

So far, I have argued that efficacy should be prioritized over fairness in the context of responding to or preventing political catastrophe. Assuming this is plausible, we must now consider what exactly precaution against political catastrophe entails.

2.5.1 General precautionary posture

We can begin to answer this question by considering some standard notions of precaution. The precautionary principle has been largely developed in the context of environmental policy and law.[67] Consider Principle 15 of the Rio Declaration:

[65] Popper 1945: 226.

[66] Notably, for Rawls (1999: 193), this restriction is necessary for preserving tolerance *and* political stability.

[67] For a number of illuminating discussions of the precautionary principle applied in environmental contexts, see Sune Holm and Daniel Steel's (2019) special issue of *Ethics, Policy, & Environment*. See also Manson 2002; Shue 2010; McKinnon 2012.

> In order to protect the environment, the precautionary approach shall be widely applied by States according to their capabilities. Where there are threats of serious or irreversible damage, lack of full scientific certainty shall not be used as a reason for postponing cost-effective measures to prevent environmental degradation.

The UN Framework Convention on Climate Change (UNFCCC) features similar language, as do the Montreal Protocol, the Maastricht Treaty, and the Third North Sea Conference.[68]

While these policy documents refer to "the" precautionary principle or approach, there is no single, universally accepted formulation.[69] (One scholar has identified nineteen versions.[70]) Nevertheless, following Jonathan Wiener, we can delineate a general "precautionary posture," which includes five elements that are common to most precautionary principles:

(1) "A threat of serious or irreversible or catastrophic risk or damage."
(2) An epistemic position that does not preclude action on the basis of scientific uncertainty about the magnitude or likelihood of a given risk.
(3) A tendency to favor "earlier measures" that "anticipate and prevent the risk" preceding harm.
(4) A tendency to favor "greater protection," though permitting degrees of stringency (from simple preventative measures to complete prohibitions).
(5) A "qualifying stance on the impacts of the precautionary measures themselves, calling for assessment of their cost-effectiveness" and "improvement over time as knowledge is gained."[71]

In sum, a precautionary posture enjoins agents to act early to prevent or attenuate major risks or damages, despite uncertainty, with the intention of securing the greatest protection possible within reasonable limits.

This way of conceptualizing precaution requires further refinement to be practicable. Some of what we have already said is useful here. We can stipulate, for instance, that the relevant notion of (1) is political catastrophe—i.e. this is what precautionary action is meant to prevent.

We have not yet addressed (2), which holds that agents should not preclude precaution on the basis of uncertainty about the likelihood of a given risk. While this condition seems reasonable, it is vague. Most pressingly, we need a clearer idea of what level of probability is necessary to warrant precautionary action. One possibility is to stipulate that risks should be "scientifically plausible" and

[68] For many more examples, see Gardiner 2006: 35; for discussion of the role of precaution in international law, see Wiener 2016.

[69] Steel 2015; Sandin and Peterson 2019.

[70] Sandin 1999.

[71] Wiener 2016: 168. See also Gardiner 2006: 36.

"reasonably likely" to occur.[72] The former requires that we understand the means by which catastrophe could occur and can explain that process to others in the language of public reason. Thus, for instance, even if Ba'al worshippers truly believe that failing to make certain sacrifices risks planetary destruction, this is a matter over which people might reasonably disagree, and without a clear way of settling this disagreement—for instance, by scientific consensus—there are no compelling grounds for taking public action against it.

The reasonably likely criterion, on the other hand, requires that the causal mechanisms capable of causing catastrophe are observably at work. This excludes very remote possibilities from our consideration, but does not rule out all low-probability or unknown-probability events. Drawing a line without raising the charge of arbitrariness is difficult here. A defensible threshold should be sensitive to potential magnitude, likelihood of occurrence, and societal aversion to risk. One candidate, suggested by the recent history of US foreign policy, is that high-impact events with a greater than 1 percent likelihood of occurrence warrant a response.[73] While this may appear overly stringent, many governments place even more demanding requirements on insurance companies in relation to their "risk of ruin": the probability that a surfeit of claims in any given year will push the insurer into bankruptcy and thus leave policyholders unprotected. For example, as a recent article in *Climatic Change* explains, the UK requires that insurance companies maintain sufficient capital holdings such that their risk of ruin does not exceed 0.5 percent within a one-year time horizon. Yet:

> In practice, insurance companies normally hold sufficient capital such that the risk of ruin is far lower than this level. Large reinsurance companies such as Munich Re and Swiss Re typically aim for a credit rating in the region of AA. An estimate of the average default probability for corporations rated AA over a one-year horizon is currently 0.02%.[74]

The article later observes that "society is currently running a larger risk" of "climate ruin" than "insurance companies are prepared or allowed to run with their own solvency."[75] This, of course, seems irrational given the far more encompassing catastrophic potential of climate change. The key point here, however, is just that, for whatever threshold of risk we adopt, events falling below it would not warrant consideration.

Henry Shue's "anti-paranoia requirement" offers additional means for refining our two risk-assessment criteria. Shue argues that we should only take precaution when (i) "we can understand the mechanisms by which [political catastrophe]

[72] Shue 2010; but cf. John 2019.
[73] This is Dick Cheney's so-called "one-percent doctrine." See Suskind 2007.
[74] Quoted directly from Bettis et al.'s creative and illuminating article (2017: 110).
[75] Bettis, et al. 2017: 116.

would happen," and (ii) "can see conditions favorable for the working of the mechanisms arising."[76] We can stipulate, then, that for a risk to be scientifically plausible, (i) must apply; while (ii) must hold for it to be considered reasonably likely to occur. This precludes action against all merely *conceivable* or *possible* threats of political catastrophe, stipulating instead that action is warranted only against risks that we understand *and* can observe developing.

2.5.2 Normative bases of precaution

Points (3) and (4) of the precautionary posture suggest two distinct normative bases for undertaking precaution: the duty not to harm others[77] and the duty to protect others from harm, respectively.

2.5.2.1 Negative duty not to harm

The former duty covers cases in which a deliberate action or omission constitutes a credible risk of significant future harm.[78] Following Lukas Meyer and Dominic Roser, we can define harm in relation to a threshold, such that "an action (or inaction) at time t_1 harms someone only if the agent thereby causes ([or] allows) this person's life to fall below some specified threshold."[79] In such cases, an agent causes or allows others to be worse off than they are "entitled to be."[80] To assess harm in this way, we must specify a particular entitlement threshold, which (again) is difficult to do without arbitrariness.[81] But the idea of political catastrophe offers one plausible candidate: a person, *P*, is harmed whenever the unnecessary actions or inexcusable omissions of another (or others) causes or allows *P* to be in a situation of extreme material scarcity, such that *P* can only meet his or her basic needs by denying another (or others) the ability to do the same.[82]

We can understand "unnecessary actions" here in two ways. First are those actions that are necessary for a satisfactory life but that are undertaken in ways that *unnecessarily* exacerbate given risks. A relevant example here is energy production: while energy is generally considered necessary for a satisfactory life, the way in which it is currently produced is unnecessarily harmful, particularly when

[76] Shue 2010: 149.

[77] The account of future harm that I will be defending here centers on basic-needs thresholds, and so avoids the non-identity problem. For a description of that problem, see Parfit 1984: ch. 16.

[78] The use of "deliberate" in this context raises important epistemic questions about the conditions necessary for an action to be counted as intentional and/or blameworthy, which I must set aside here (for interesting discussion, however, see e.g. Lichtenberg 2010; Broome 2021b). Deliberate should be taken here simply to refer to actions undertaken in awareness, or inexcusable ignorance, of potential ill consequences.

[79] Meyer and Roser 2009: 228; cf. Shiffrin 1999.

[80] Meyer and Roser 2009: 229.

[81] Arneson 1999, 2000. But see also Pölzler's (2021) helpful discussion.

[82] Much like the thresholds put forward by Meyer and Roser 2009, Wolf 2009, and Rendall 2011, this one centers on the ability to satisfy basic needs.

cost-effective carbon-neutral technologies are available. Second are actions that contribute to a given risk and yet are unnecessary for a satisfactory life.[83] For example, consuming animal products appears unnecessary for (and often even detrimental to) a healthy and satisfying life,[84] at least for most people in the West.[85] Nevertheless, the livestock sector has exploded in the last fifty years, particularly in the West, and is now "responsible for 18 percent of greenhouse gas emissions"—more than all the world's trains, planes, cars, boats, and other transport vehicles combined.[86]

The duty not to harm also proscribes "inexcusable omissions." An omission, or failure to act, is excusable if the party failing to act is (i) ignorant of the consequences and (ii) lacks the means to overcome that ignorance. Otherwise, it is inexcusable. Thus it may be wrong to hold generations before, say, 1990 accountable for harming present and future generations with GHG emissions, given the lack of broad scientific consensus about the causes and effects of climate change up to that point.[87] Failing to take preventative action today, however, is inexcusable, as this ignorance no longer exists, or, where it does, the means of overcoming it are readily available.

2.5.2.2 Positive duty to protect

The second normative basis of precaution follows from the positive duty to protect others from harm.[88] Much like the closely related duty to rescue, the content and scope of the duty to protect are controversial. In standard formulations, however, the duty to protect holds that when one is able to prevent harm from befalling another (or others) at reasonable cost to oneself, then one ought to do so. Here again harm may be defined in relation to a threshold, such that others (e.g. future

[83] I set aside the question of what constitutes a sufficiently satisfactory life, but see e.g. Sen 1984; Rawls 1999: 54–5, 78ff.; Shue 1999: 542; Nussbaum 2000. See also Robeyns 2017 for a defense of the pertinent claim that "it is not morally permissible to have more resources than are needed to fully flourish in life."

[84] On this, see e.g.; Ripple, et al. 2014; Stanaway, et al. 2018; Afshin, et al. 2019; IPCC 2022a: ch. 7, p. 1075 or, for a more general and accessible treatment, Greger and Stone 2016.

[85] One might object that certain gustatory delights, particularly those of consuming animals, are necessary for a satisfactory life. I do not need to deny this to establish that there is an obligation to exclude animal products from one's diet, so long as affordable alternatives are available that do not unnecessarily subject other persons to political catastrophe. To counter this, one must establish that *only* animal products can provide gustatory pleasure of the sort that constitutes a satisfactory life, which is unlikely. Moreover, one must argue that those delights are more significant to a satisfactory life than the basic goods that are threatened by animal agriculture (e.g. access to clean water and food, a stable climate). Even if this could be shown in the aggregate—i.e. that the total pleasure received from consuming animal products outweighs the suffering animal agriculture causes and risks—those committed to non-utilitarian justice theories must also establish how the particular pleasure of consuming animals justifies the harm animal agriculture inflicts on others (including nonhuman others).

[86] Steinfeld, et al. 2006: xxi. See also Wellesley, et al. 2015: vii; Herrero, et al. 2016; Clark, et al. 2020; Harwatt, et al. 2020; Xu, et al. 2021; Ripple, et al. 2022; Scarborough, et al. 2023.

[87] On excusable ignorance, see e.g. Kingston 2014; Mittiga 2019a: 165ff.; García-Portela 2020; Truccone-Borgogno 2023.

[88] Gardiner 2006.

people) are harmed whenever their quality of life falls below a certain level. Unlike the negative duty not to harm, however, the positive duty to protect holds whether or not an agent is responsible for the potential harm. Thus, assuming climate change threatens political catastrophe (a point I defend below), this duty would require us to undertake precautionary action *even if* we were not personally or principally responsible for causing the risk.

While the duty not to harm is absolute—only ignorance or necessity can excuse deviations—the duty to protect must have limits or else it will be overly demanding.[89] This is intimated by elements (4) and (5) of the precautionary posture, which stipulate that, while we should favor a greater degree of protection over a lesser one, considerations of cost-effectiveness may limit what actions we take. This again is vague, however. How stringent should we be in protecting others, particularly against risks of political catastrophe?

A point from the contractualist priority argument laid out above (section 2.4.2.1) is relevant here. Recall that, from an impartial perspective, a rational limit to our precautionary duty is given by political catastrophe itself. In allocating burdens to prevent political catastrophe, none should be so great that they themselves result in extreme material scarcity. This prevents precaution from becoming self-defeating and thus preempts the so-called "black-hole problem," which affects principles that prioritize the prevention of worst-case scenarios.[90] The concern is that strictly prioritizing the prevention of such scenarios would quickly exhaust society's resources. For this reason, many prefer principles that minimize aggregate or average deprivation.[91] Although my argument gives strict priority to preventing political catastrophe, it hedges against the black-hole problem by precluding the complete material sacrifice of any group for the sake of others.

2.5.3 The pluralist precautionary approach

In light of these considerations, we can formulate a more precise account of our precautionary climate duties as follows:

(1) GENERAL PRECAUTIONARY DUTY: All agents have a general duty to prevent scientifically plausible, reasonably likely threats of political catastrophe where possible. This implies two correlate duties:

[89] On this, see e.g. Munthe 2019.

[90] Wolf 2009: 356–7. Note that Wolf considers this problem in application to prioritarian intergenerational equity principles, but the general concern also applies here. For a statement of a similar concern with regard to precautionary principles, see Sunstein 2006: 869ff.; Munthe 2019. See also Bostrom 2013.

[91] See e.g. Sunstein 2006: 893; Meyer and Roser 2009; Wolf 2009: 357; Rendall 2011: 888.

(a) DUTY NOT TO HARM: Agents should not harm others. Harm occurs whenever some agents, by their unnecessary actions or inexcusable omissions, cause or allow political catastrophe to befall others.

(b) DUTY TO PROTECT: Agents should take all reasonable measures to protect others from harm in the form of political catastrophe.

  i. REASONABLENESS PROVISO: The limit of reasonability in (1b) is given by political catastrophe itself: precautionary burdens cannot be so great as to themselves cause politically catastrophic conditions.

And to this, we should add the following secondary principle:

(2) FAIRNESS PRINCIPLE: Agents should ensure the burdens of precaution are distributed fairly to the fullest extent compatible with satisfying (1a) and (1b).[92]

Taken together, these principles provide an account of precautionary duty that balances moral and realist political commitments. Accordingly, we can hereafter refer to it as the "pluralist precautionary approach" (or PPA).

While novel in structure and formulation, the PPA has practical implications similar to those of John Broome's well-known "efficiency without sacrifice" (EWS) proposal.[93] For Broome, "the very most important thing about climate change" is that it "can be solved without anyone making any sacrifice."[94] In other words, "fixing climate change" is possible, he contends, without imposing (uncompensated) economic costs on current generations; recognizing this effectively neutralizes the greatest political objection to immediate action.[95] Notably, this sounds like a version of the intergenerational-justice argument described in section 2.2. And indeed Broome accepts one of the basic premises of that argument: that "[u]nless climate change turns out to be much more damaging than expected, economic growth will continue, so that ... future people will be better off than we are"; consequently, "[g]iving them more benefits at our expense would be benefitting the rich at the expense of the poor."[96] Critically, however, Broome does not present EWS as a way of preserving the priority of fairness; on the contrary, he regards it as a significant moral compromise, in that EWS requires accepting some amount of unfairness for the sake of motivating people to implement action now and thereby prevent future harm.

92 I set aside here the question of which principle or standard of fairness (e.g. sufficientarian, prioritarian, strict egalitarian) we should employ after the general precautionary duty has been satisfied.

93 Broome 2012: esp. 44–8.

94 Broome 2010, although he attributes this view to Foley (2007) and Stern (2010a).

95 Broome 2012: 47.

96 Broome 2012: 45–6.

Far better than EWS, morally speaking, would be "efficiency *with* sacrifice," Broome argues. This is so for a number of reasons. Perhaps most significantly, EWS essentially amounts to a "bribe" paid *by* future people—who, through no fault of their own, are at risk of grave harm—*to* those alive today (above all, the global rich)—who will continue causing harm (via emissions) unless they are compensated for acting otherwise.[97] This suggests serious moral corruption: current generations should not require a bribe from their potential victims to fulfill their basic moral duty not to harm others.

Moreover, Broome argues there are strong consequentialist reasons for preferring policies that require present sacrifices. As he explains, while "[o]ther things being equal, it would be a bad idea to benefit the rich at the expense of the poor," "if the total of benefits can be greatly increased by doing so," as he argues would be the case following an efficiency-*with*-sacrifice plan, "it may be a good idea."[98]

In short, then, Broome believes that the moral balance of reasons is tipped unequivocally toward policies that would involve present action *and* sacrifice. Nevertheless, he endorses EWS as a *strategically* more viable position, and one that is at least morally superior to "business as usual" climate politics: "Aiming for *efficiency with sacrifice* rather than *efficiency without sacrifice* is to encumber the task of fixing climate change with the much broader task of improving the distribution of resources between generations."[99] In other words, for Broome, efficaciously responding to the threat of climate change is more important than satisfying the full requirements of distributive fairness, and thus should the former become considerably more likely to succeed if we were to set aside the latter, then we have good reason to do that.[100] Moreover, at least with EWS, Broome argues, we have something that "is better for some people and worse for no-one."[101] To insist on anything more than that would be to "make the best the enemy of the good."[102]

The pluralist precautionary approach (again, PPA) shares with Broome's view the basic idea that undertaking effective action must not be made conditional on satisfying the full demands of distributive fairness. In other words, however regrettable it may be, if the only viable options we have for responding to the perilous threat of climate catastrophe are in some way unfair, we must pursue them nevertheless. Unlike Broome's view, however, the PPA does not locate the reason for this outside of justice—i.e. by appeal to strategic or prudential considerations alone.

[97] Broome 2012: 46, 43–4.

[98] Broome 2012: 45, 44.

[99] Broome 2012: 47.

[100] Notably, in recent work, Broome has made even more explicit his endorsement of the view that some amount of injustice should be tolerated if doing so is necessary to address the crisis, and that we should not insist on moral solutions or moral reformation. See e.g. Broome 2016: 21–33, and Broome 2021a.

[101] Broome 2012: 45.

[102] Broome 2012: 47.

Rather, according to the PPA, unfairness in the context of preventing climate catastrophe is permissible for the sake of preserving the conditions that make fairness possible; thus, we act out of a concern for justice, not in spite of it.

It does not follow from this, however, that the PPA is incompatible with Broome's EWS proposal. Quite the opposite: should it be the case that an EWS policy were able to prevent climate catastrophe and be the fairest option we have among comparably effective proposals, then the PPA would readily endorse it. The key difference, however, is that the PPA is able to offer a deeper, more explicitly moral defense of why this would be an acceptable approach, despite its unfairness.

2.6 Objections to the pluralist precautionary approach

In what remains of the chapter, I consider several potential objections: (1) that climate change will not actually be politically catastrophic; (2) that, on various plausible accounts of intergenerational justice, the PPA is superfluous and so unnecessary; (3) that the PPA has ambiguous policy prescriptions; and (4) that the PPA would legitimate the use of dangerously illiberal and undemocratic emergency powers.

2.6.1 Climate change will not be politically catastrophic

The first objection charges that although climate change may cause serious future damage, it is unlikely to precipitate political catastrophe; thus there is no need to undertake extensive precautionary action, particularly if doing so comes at the cost of justice.

This is an empirical question that we can settle by examining the evidence. I argue that climate change threatens political catastrophe both *globally* and *regionally*. The former pertains to climatic changes and events that are worldwide in scope, totalizing in effect, and irreversible, though unlikely to occur before 2100. The latter pertains to less severe, but more immediate, climatic changes and events capable of upending the objective circumstances of justice in vulnerable states or regions.

2.6.1.1 Globally politically catastrophic climate change

Climate change threatens global political catastrophe in two main ways: via (i) linear increases in average surface temperatures (in excess of, say, 4°C);[103] and (ii)

[103] As Catriona McKinnon (2012: 60–1) astutely observed, more than ten years before this book was published: "The net result of a 5°C increase would be worldwide famine, severe conflict over water in the world's desertified regions, and extreme territorial insecurity for those living in the dwindling

severe climate events—especially "tipping points"—capable of abruptly altering weather patterns or ecosystems in devastating ways.

[*i*] As noted above, the IPCC reports that a doubling of CO_2e will "likely" result in a net increase of 1.4°C to 4.5°C in mean temperatures over preindustrial levels. The IPCC defines "likely" as having a 66 percent chance, meaning that *if* we stabilize CO_2e at around 550 ppm—an ambitious goal, given that the atmospheric stock of CO_2e is already in excess of 400 ppm and continuously rising—there remains a 34 percent chance that global warming will fall outside the 1.4–4.5°C range. Note, however, that it is considerably more likely that temperature increases will exceed 4.5°C than fall below 1.4°C.[104] And on one estimate, there is as much as "a 10% chance of eventual temperatures exceeding 6°C."[105]

Although six degrees of warming may sound innocuous, it is enough to extinguish or severely endanger much of the planet's plant and animal life—much as a fever of 6°C above normal body temperature is enough to kill a human being.[106] According to Nicholas Stern, with just a 5°C rise, "[h]uman life would probably become difficult or impossible in many regions that are currently heavily populated, thus necessitating large population movements [... which] often bring major conflict."[107]

The links between global warming and political catastrophe are manifold.[108] Higher temperatures mean increasingly volatile weather patterns; rising sea levels; higher rates of infectious diseases;[109] massive displacement and waves of "climate refugees";[110] more frequent drought and flooding (and thus more frequent crop failure); and extensive water scarcity.[111] Stern is not alone in predicting conflict—the White House, the US Department of Defense, NATO, etc.[112] all emphasize the possibility of intra- and interstate violence arising from global warming. As Martin Weitzman describes, "[t]he massive unrest and uncontainable pressures [higher temperatures] might bring to bear on the world's human population are almost unimaginable."[113]

[*ii*] Another path to global political catastrophe involves surpassing "tipping points"—i.e. "thresholds beyond which major changes occur that may be self-reinforcing and are likely ... irreversible over relevant time scales."[114] These

inhabitable areas as an archipelago of refuges is put under increasing pressure by hungry and desperate people."

104 Indicative of a "fat-tailed distribution" (Wagner and Weitzman 2016: 49–51).

105 Wagner and Weitzman 2016: 53.

106 Gernot Wagner made this analogy during a University of Virginia Environmental Humanities Colloquium presentation on December 1st, 2015.

107 Stern 2010b: 44.

108 For further discussion of this point, see Chapter 1.

109 Romanello, et al. 2021.

110 Eckersley 2015; Thalheimer, et al. 2023.

111 Posner and Weisbach 2010: 29; Singer 2010: 183; Stern 2010b: 44.

112 US Department of Defense 2015; Nuccitelli 2017; The White House 2022.

113 Weitzman 2012: 232.

114 Furman, et al. 2014: 20. See also McKinnon 2012: 48ff.

thresholds are typically defined by specific levels of warming; it is impossible, however, to predict precisely how much temperatures would have to increase to set off a given tipping point. In other words, we may pass critical warming thresholds, and thereby set off a tipping point, suddenly and without warning. According to a National Research Council report, one tipping point may have already been crossed: the destabilization of the Greenlandic and West Antarctic Ice Shelves.[115] This threatens to greatly accelerate sea-level rise and, worse, shut down the North Atlantic thermohaline circuit, which plays a crucial role in moderating temperatures in Europe.[116] When the thermohaline circuit last stopped, global temperatures dropped 5°C within ten years, causing icebergs to extend to Portugal, which locked in enough water to precipitate a severe global drought.[117]

Other potential tipping points, like the release of methane stores in the Northern hemisphere's thawing permafrost, risk "runaway" global warming.[118] In general, events like these could have enormous ramifications for well-being and political order—which is to say, crossing tipping points credibly risks future global political catastrophe. The only viable way to avert tipping points is by reaching carbon neutrality as quickly as possible.

2.6.1.2 Regionally politically catastrophic climate change

Catastrophic linear temperature increases and many of the most disruptive tipping points may not occur until the end of the twenty-first century. In the meantime, however, climate change poses a credible risk of political catastrophe on a smaller scale, in particular regions and states.[119] Economists and policymakers have paid far less attention to this possibility, perhaps because the changes and events culminating in regionally politically catastrophic climate change are less sweeping and may be (partly) remediable through international action.

To understand how climate changes risk regional political catastrophe, note first that when scientists speak of temperature increases they are typically referring to global averages. This can obscure the fact that even minor increases in global mean temperature can entail considerable increases for certain areas.[120] Moreover, tipping points can be localized—increases of just 1–2°C can abruptly impact regional weather patterns, resulting in protracted droughts, topsoil aridification, etc.[121] Warming in the Pacific, for instance, has already led to considerably more frequent and severe cyclones in east Asia.

[115] National Research Council 2013. This has been further supported in more recent research; see e.g. Feldmann and Levermann 2015; Khazendar, et al. 2016.

[116] Rahmstorf, et al. 2015; Castellana, et al. 2019; Ditlevsen and Ditlevsen 2023.

[117] Gardiner 2004: 562–3.

[118] See Whiteman, et al. 2013 (cited in Wagner and Weitzman 2016: 185).

[119] For a brief but helpful discussion of this possibility, see Steel, et al. 2022.

[120] Posner and Weisbach 2010: 18–20. The most recent IPCC reports (particularly IPCC 2021a and IPCC 2022a) provide more technical discussion of variations in regional impacts.

[121] See e.g. IPCC 2021b: 16.

Regional climate change does not necessarily entail political catastrophe. Rather, catastrophe will occur only when affected states' or regions' anticipatory and response capabilities are overwhelmed and international assistance is lacking. This suggests two areas on which to focus precautionary action beyond abatement: (i) ensuring that affected states or regions are prepared for climatic changes by providing the necessary resources to implement effective adaptation measures;[122] and (ii) providing assistance or compensation to attenuate the effects of unavoidable loss and damage.[123]

2.6.2 Precaution is superfluous

A second objection holds that sufficientarian accounts of intergenerational justice—like Rawls's just-savings principle or Locke's more inchoate "enough-and-as-good" proviso—already adequately insure against political catastrophe by requiring that contemporaries not deprive future people of adequate material resources. If this is right, then there is no need for the pluralist precautionary approach (PPA), or perhaps any other theory of precautionary duty.

This objection misses the difficulties sufficientarian accounts face in the following, pertinent case: when contemporaries act in such a way that enriches future people (in pecuniary terms) but also severely depletes or degrades the natural environment.[124] Because accounts like Rawls's and Locke's assume fungibility between natural and artificial material resources, even irreversible environmental damage is regarded as compensable. Accordingly, even if future generations are forced to adapt to severe environmental austerity, these accounts assume that their greater wealth and productivity should make this, on balance, acceptable to them.[125] (Hence Brian Barry's observation that, on accounts like these, a relevant question becomes: what amount of manufactured plastic trees, astroturf, and singing electronic birds can compensate for the loss of real trees, real grass, and real birds?[126])

In this sense, Rawlsian or Lockean intergenerational sufficientarianism appears to favor promoting economic growth over preventing environmental loss. It is by no means clear, however, that adapting to unmitigated climate change is (or at least will remain) a viable option for future generations, even granting extensive

[122] On this point, see Posner and Weisbach 2010: 22.

[123] "Food aid plays an important role in providing food security and saving lives after climate disasters" (IPCC 2018a: 5.2.5). See also Wallimann-Helmer 2015; Wündisch 2019; Capisani 2020.

[124] Notably, according to the IPCC, the least ambitious climate policy pathways—i.e. those that would permit the highest degree of warming—are also considered the most conducive to GDP growth (see IPCC 2022b: C.12.2, p. 41).

[125] Economists like Nordhaus and Boyer explicitly endorse this view, as do some philosophers (e.g. Heath 2021).

[126] Barry 1999: 102.

"wealth" accumulation. Insisting otherwise requires the implausible assumption that no amount of ecological devastation can exceed our ability to adapt. This reproduces the problematic view challenged earlier—that the condition of moderate scarcity is invulnerable to human action—with the sole difference that artificial goods are construed here as suitable replacements for natural goods. The objection thus fails to take seriously the risk of political catastrophe. So long as credible doubt exists over whether endless adaptation is possible, precaution remains necessary.

2.6.3 The PPA is prescriptively ambiguous

A third objection holds that it is difficult to identify minimum requirements for preserving the condition of moderate scarcity, and this indeterminacy should lead us to reject the PPA, or at least exclude it from practical deliberation.

One way to respond is to assert a simple baseline: that justice requires material conditions to be such that all individuals can (continue to) meet their basic needs without having to deny others the ability to do the same.[127] While this may be vague, defining a more exact threshold is not essential. It suffices to fix on credible risks of political catastrophe (like those posed by climate change), and structure our action so as to reduce the likelihood and/or potential magnitude of those risks as far as we are reasonably able to do, which is precisely what the PPA prescribes.

2.6.4 The PPA sanctions dangerous state power

The final objection holds that subordinating ordinary moral constraints to efficacy seems to entail vesting states with dangerous emergency powers, thus opening the door for authoritarian abrogations of rights, diminutions of democracy, or other such abuses. For this reason, the PPA should be rejected.

Responding to this concern, and defining the justification and proper scope of climate emergency powers, comprises the focus of Chapter 3. It will be helpful to consider a few points here, however. First, there is no reason to suspect that undertaking precautionary action against politically catastrophic climate change would require sacrificing any of the most intimate individual freedoms and rights—e.g. freedoms of conscience or worship, or the right to peaceably assemble. Moreover, if fulfilling our precautionary duties were to impinge on individual rights, it must do so only to the extent necessary for alleviating or preventing political catastrophe.

Of course, any potential limitation of rights may rightly be considered disconcerting; it is important to recognize, however, that, should such limitations indeed

[127] Cf. Rawls 2005: 7; Meyer and Roser 2009: 225n.11; Wolf 2009.

be necessary, this would not constitute a break from ordinary politics, inside or outside of emergency scenarios.[128] Consider the possibility of prohibiting climate disinformation campaigns like those that have been carried out to great effect for decades by the fossil-fuel industry.[129] This would undeniably constitute censorship, and so represent a limit on free expression. Yet freedom of expression has never been absolute: even the most liberal thinkers, like J.S. Mill, make exceptions for harm-inducing speech,[130] and even the most liberal states maintain statutes against libel, hate speech, inciting public riots or violence, and so on.[131] Likewise, in the course of fulfilling our precautionary duties, we may think it necessary to curtail certain property rights, for example by prohibiting fossil-fuel producers from developing the oil assets they already own, or by severely limiting the scope and duration of green technology patents so as to encourage their propagation. These measures, again, would not mark a significant departure from normal practice, particularly during emergencies: it is often thought legitimate, for instance, to seize and redistribute the stocks of food suppliers during a famine, or to nationalize and repurpose factories during a war.[132]

The same basic point holds with respect to the possibility of imposing limits on democratic institutions or practices. There are no plausible grounds for thinking that precaution would require any of the most serious antidemocratic measures, such as suspending elections, disenfranchising voters, or instituting fully dictatorial regimes. (Again, I take this up in far greater detail in Chapter 3.) Yet a commitment to precaution may lead us to consider implementing certain limits on what democratic decisions can authorize, much as ordinary constitutional restrictions already do. For instance, we might stipulate that, just as no democratic public or representative government can authorize enslavement, neither can they enact policies that pose a scientifically plausible, reasonably likely risk of political catastrophe. We might also consider barring citizens with deep ties to the fossil-fuel industry from holding certain public offices, particularly those who have a history of public climate denialism. Again though, such measures would not constitute a radical break with existing practices in many places. In postwar Germany and

[128] Nomi Lazar (2009) offers a compelling argument for the continuity of ordinary and emergency government—i.e. for regarding emergency powers as "an extension of everyday practice and everyday values." To simplify considerably, her essential claim is that, while many regard emergency situations as exceptional insofar as they involve derogations of basic rights, this misses the fact that similar derogations occur all of the time, as part of the routine enforcement of law and maintenance of public order. In this sense, the difference between "everyday" and emergency government is ultimately one "of degree and not of kind" (Lazar 2009: 100).

[129] See Oreskes and Conway 2011.

[130] See the famous discussion of corn-dealers in chapter III of his *On Liberty* (Mill 1998). See also Lazar 2009: 91.

[131] For a helpful general discussion, see van Mill 2021.

[132] Another pertinent example is the relaxation of intellectual property rights or patent restrictions to vaccines or other essential medicines during a pandemic, as has been proposed (albeit in a limited way) during the COVID-19 pandemic (Furlong 2022).

Austria, for instance, former Nazi Party members were barred from holding public offices; and so too are convicted felons in Texas[133] today.[134]

These examples suggest a second line of response, concerning public reason and the weight of moral duties. Many political obligation theorists maintain that duties only hold in a *pro tanto* sense—i.e. that they may be justifiably outweighed in certain (exigent) circumstances. But these theorists also carefully stipulate that, while exigency may *override* duty, it never *erases* it. For this reason, an explanation is owed to adversely affected parties after the fact. Crucially, this suggests a requirement of accountability. Such a requirement would mean that those (e.g. current generations) who are to bear the heaviest burdens of preventing politically catastrophic climate change are owed an explanation, assuming that overall distribution of burdens is in some plausible way unfair. Such explanations exist—as this chapter is, in part, meant to show. But in addition to general explanations, specific explanations for particular policies are necessary. These should specify clearly who is responsible for acting, why, when, and how. Accountability can thus protect against ineffective, unnecessary, or inordinately burdensome policies.

2.7 Conclusion

This chapter has considered under what conditions, and to what extent, we can justifiably prioritize efficacy over fairness, particularly in the context of the climate crisis. I began by surveying two main responses to these questions. The first, prominent in the intergenerational-justice literature, attempts to avert any trade-off by separating the costs of climate action from the actions themselves, and stipulating that the former be deferred to (likely) richer future generations. With this, the most efficacious distribution of climate-action burdens (i.e. one that imposes strenuous requirements on earlier, poorer generations) remains fair. As I showed, however, this proposal falls short in a number of respects. For one, there is great uncertainty about the future effects of climate change and the actual economic situations of future people. More problematically, it is not clear that costs can actually be transferred between generations; and even if some could, many more (like the costs of lost employment) simply cannot, meaning that some injustice will remain.

The second response, advanced by Simon Caney, takes as its starting point the urgent need to prevent climate catastrophe. It then proceeds to elucidate a set of conditions under which we would be justified in prioritizing the prevention of harm over ensuring that the burdens of cooperation are fairly distributed. While

[133] See Section 141.001 of the Texas Election Code (https://statutes.capitol.texas.gov/Docs/EL/htm/EL.141.htm).

[134] This measure may seem more obvious if states begin to heed the call of certain ethicists, like Catriona McKinnon (2017), and treat "postericide"—i.e. "intentional or reckless systematic conduct that is fit to bring about near human extinction"—as a crime against humanity.

Caney's account is in many ways compelling, extending it to cover core climate actions—like reducing emissions or supporting adaptation—proves problematic. Moreover, several of the conditions Caney sets out are vague (e.g. how likely must a risk of future harm be before action is justified?; which agents should act when all are not needed, and why?).

I thus proposed an alternative: the pluralist precautionary approach (PPA). Like Caney's account, the PPA starts with the assumption that preventing catastrophe is essential. Yet it focuses on a unique form of catastrophe: what I call political catastrophe, which, again, occurs whenever material scarcity becomes so extreme and enduring that agents are unable to satisfy their most basic needs without denying another or others the ability to do the same. Under such conditions, realizing or sustaining fair terms of social cooperation, and thus political stability, becomes impossible.

When faced with political catastrophe, I argue that agents are justified in prioritizing precautionary efficacy over fairness. Crucially, though, this is not for reasons external to justice but for the sake of justice itself; my essential claim is that we may relax or suspend *particular* principles and rules of justice if doing so is necessary for preserving the material conditions that make justice *in general* possible. In practice, this means subordinating considerations of fairness to those of precautionary efficacy.

This argument has direct relevance to the problem of climate change, which, if left unaddressed, promises political catastrophe on a global scale. Regrettably, the only way to prevent catastrophe at this late stage may involve imposing unfair burdens on some for the benefit of others. This is especially clear in the intergenerational context: precautionary climate action is likely to succeed only if current, (presumably) poorer generations make potentially substantial sacrifices now for the benefit of (presumably) wealthier future people. But the same dynamic may hold true intragenerationally, that is, among states today: it may, for instance, be impossible to avert catastrophe without imposing unfair costs or developmental delays on high-emitting but relatively poorer states like China and India—burdens for which they may never be fully compensated.

None of this is to say that fairness should be totally evacuated from the political calculus. Quite the opposite: when weighing comparably effective schemes of action, we should always prefer the fairest option. Nevertheless, if a trade-off between fairness and efficacy *is* necessary, then the latter must take priority. For political catastrophe is inhospitable to everything that we value, fairness included. A fundamental end of government—and one of the deepest sources of its legitimacy[135]—must be to prevent political catastrophe from arising; for should it fail at this, there can be no justice, no rights, no democracy, and no stability, at least for long.

[135] As I argue in Chapter 3.

3
Political Legitimacy, Authoritarianism, and Climate Change

... the fundamental law of nature being the preservation of mankind, no human sanction can be good, or valid against it.

John Locke, *Two Treatises*, ch. 11, par. 135

3.1 Introduction

As the climate crisis deepens,[1] one can find a cautious but growing chorus of praise for "authoritarian environmentalism."[2] This mode of governance, typically associated with China, is often juxtaposed to the so-called "democratic environmentalism" of wealthy post-industrial states like the US, Australia, Germany, and Japan. The essential idea behind these encomiums is that, while authoritarianism is in general lamentable, having a government unencumbered by democratic procedures or constitutional limits on power could be advantageous when it comes to implementing urgently needed climate action.

It is ultimately an empirical question whether authoritarian governance is better able to realize desired environmental outcomes, and, if so, why and to what extent.[3] Yet it is undeniable that nearly all wealthy democratic states have failed to respond adequately to the climate crisis. By contrast, various less-affluent authoritarian regimes have been successful in implementing stringent climate policies, and several are now considered global leaders in the production and installation of "solar panels, wind turbines, batteries, and [the adoption of] electric vehicles."[4] But while questions about the relative efficacy of democratic and authoritarian *regimes* are highly relevant, they do not comprise my focus here. Rather,

[1] An earlier, slightly truncated version of this chapter was published in the *American Political Science Review*, vol. 116, no. 3, pp. 998–1011 (Mittiga 2022). I thank the journal for allowing this material to be reprinted here.

[2] The terminology varies between authors; some refer to "authoritarian environmentalism," others to "environmental authoritarianism," "eco-authoritarianism," or, less commonly, "green authoritarianism." See e.g. Beeson 2010; Gilley 2012; Han 2015; 2017.

[3] According to Stevenson and Dryzek (2014), there is at least some evidence that more deliberatively democratic governments perform better. But cf. Bell 2016; Povitkina 2018; von Stein 2022.

[4] Kerry and Khanna 2019.

Climate Change as Political Catastrophe. Ross Mittiga, Oxford University Press.
 DOI: 10.1093/oso/9780192868879.003.0003

I am interested in determining under what conditions authoritarian forms of climate governance may be considered legitimate, including in otherwise liberal-democratic states; and, more broadly, how governments' responses to climate change impact normative assessments of their political legitimacy.

To date, political theorists have not thoroughly examined these questions.[5] They have been explored in some popular media, however. In the television adaptation of Margaret Atwood's *The Handmaid's Tale*, for instance, a commander of the "Republic of Gilead"—a brutal totalitarian theocracy, responsible for overthrowing the US government—boasts to foreign ambassadors that his government has "reduced carbon emissions by 78% in just three years."[6] His point is not technical but normative: for the commander, responsible climate action stands as a source of legitimacy. Of course, no environmental policy—however effective or direly needed—is sufficient to legitimate a regime that commits horrifying violations of human rights like the chattel slavery of women. Yet it is difficult to deny that *some* relationship exists between climate action and legitimacy.

Understanding this relationship requires thinking more carefully about what constitutes political legitimacy in general. Some—like Thomas Hobbes and various contemporary realists—argue that it depends, minimally, on a government's ability to ensure the safety and security of those it governs. Others—particularly those within the democratic and liberal traditions—adopt a moralized view, claiming that legitimacy requires, e.g. consent, free and fair elections, equal representation, protections of individual rights, social justice, or, most often, some admixture of these factors. While it is tempting to regard the former "realist" view and the latter "moralist" view as representing competing answers to the same question ("what constitutes political legitimacy?"), I argue that they are typically complementary, their demands only coming into conflict in dire conditions. Some thinkers—most notably, Bernard Williams—have identified this dual dynamic;[7] however, the precise relationship between the realist and the moralist conceptions of legitimacy remains vague, particularly in the context of climate change.

I begin addressing this ambiguity in section 3.2 by distinguishing between what I call "foundational legitimacy" (FL) and "contingent legitimacy" (CL). Crudely put, FL requires that citizens' essential safety needs are met; it is *foundational* in the sense that no government can be legitimate unless it satisfies this demand. CL,

[5] There are some partial exceptions. Mulgan (2014), for instance, examines the incoherency of a liberal-democratic politics in a future world marked by catastrophic climate change. Some older work, like that of Ophuls (1973) or Walker (1988), also explores the tension between maintaining liberal democracy and confronting environmental disaster; they do not, however, approach this issue through the lens of legitimacy.

[6] Season 1, episode 6.

[7] Nomi Lazar's (2009: 88) illuminating work on states of emergency in liberal democracies also seems to affirm something like this view: "liberal values constitute quotidian ethics animating the day-to-day functions of a political order, while existential ethics invoke different principles for the preservation of the liberal democratic order." On the difference between quotidian and existential ethics, see Lazar 2009: 22ff.

on the other hand, requires that the power used by the government—to secure and maintain FL, along with other ends—be, in principle, "acceptable" to all those who are subjected to it. What counts as acceptable is, of course, a matter of debate, which helps to explain why so much of the literature on legitimacy unfolds on this level. I claim, however, that a core feature of all CL standards is that they shift over time, across cultures, and in light of new circumstances, and that it is in this sense that they are *contingent.*

In section 3.3, I examine the relationship between FL and CL. I argue that under reasonably favorable conditions, FL functions as something of an unacknowledged background condition of CL. In moments of great exigency and upheaval, however, tensions between FL and CL can and often do arise. In such cases, I claim that preserving or restoring FL should take priority over satisfying any particular demand of CL. In other words, from the perspective of political legitimacy, ensuring safety and security may at times justify relaxing or suspending[8] strict adherence to certain democratic processes or individual rights.[9]

Although the idea of such trade-offs will strike many as troubling, we should note that they already constitute a nearly ubiquitous—if often only implicit—element of contemporary political practice.[10] In times of war, for instance, authoritarian impositions of power, including those that curtail democratic processes or basic rights, are often thought legitimate to the extent they are necessary for protecting citizens and restoring normal conditions. Likewise, as those who have survived COVID-19 can attest, during a health emergency, severe and enduring limitations of rights to free movement, association, and speech can become legitimate techniques of government, even in robustly liberal-democratic states. As these examples suggest, in crisis moments, political legitimacy may not only be compatible with authoritarian governance but *require* it. Conversely, stringent adherence to liberal-democratic constraints may diminish legitimacy insofar as this inhibits effectively responding to credible security threats.

Again, many may regard the prospect of such moral compromise as deeply problematic. To be clear, though, the argument presented here should not be understood as an endorsement of authoritarianism but rather as a *warning*: should we wish to avoid legitimating authoritarian governance, we must do all we can to prevent emergencies from arising that can only be solved with such means.

Lamentably, this is not always possible: some emergencies arise that are impossible to predict in advance, while others may be foreseen but simply not addressed in time. The latter seems to be true of climate change, which, as I argue in section 3.4, is (or may very soon become) sufficiently grave and disruptive as

[8] See Adams and Mittiga (2021: 2238n.3) for a description of what is meant by "relaxing or suspending."

[9] This is *not* to say that we should always be willing to accept abridgments of rights or democracy for safety, as I explain in section 3.3.

[10] See e.g. Rossiter 1948; Agamben 2005: ch. 1; Lazar 2009: esp. ch. 4.

to jeopardize FL. Therein lies the allure of authoritarian environmentalism: for if, as many now contend, liberal-democratic norms, principles, and institutions impede urgently needed climate action, then legitimacy may permit—or even require—those constraints to be relaxed or abandoned.

While we may wish to resist the view that the factors most commonly associated with CL—like respect for democracy or basic rights—could ever meaningfully inhibit climate action, examples of precisely this are manifold. Democratic publics have, on multiple occasions, defeated (via referendum or protest) even modest carbon taxes.[11] Free speech rights in many countries have made regulating harmful climate denial and disinformation campaigns virtually impossible.[12] Likewise, the primacy of individual autonomy has at times rendered even minor interventions (e.g. around lightbulbs,[13] fuel efficiency standards,[14] or diets[15]) extremely contentious, and more ambitious policies (like population control) totally unthinkable. Given this, liberal-democratic governments (and theorists) must confront the bleak possibility that responding to the existential threat of climate change at this late stage may (soon) require relaxing or suspending adherence to some of the most widely shared CL standards, and embracing authoritarian power.

I argue further, in section 3.5, that as climate change is likely to be disruptive for a long time, it may also cause a more permanent reconfiguration of CL standards. That is, the climate crisis may not just lead to, e.g. temporary and localized suspensions of democratic processes or individual rights, but precipitate a more substantial and enduring shift in what counts as an "acceptable" use of political power. While there is space for concern about such changes, they may not be all for the worse. Imagine, for instance, if democratic representation came to be understood in intergenerational terms, such that only those governments that awarded formal standing to future people were considered legitimate.[16] Or if individual human rights were thought to be predicated on, and therefore limited by, a more basic biotic right to continued existence, shared by all living beings.[17] Section 3.5 limns three potential CL factors in this spirit, all of which, I claim, are already

[11] Consider Washington state's rejection of Initiative 1631, which would have implemented a carbon tax (Roberts 2018); or the "gilets jaunes" protests against increased fuel taxes in France (BBC News 2018b).

[12] Consider, for instance, the resistance to the exceedingly modest recent proposal to limit advertising for carbon-intensive personal vehicles in the UK (Laville 2020).

[13] In 2007, a rather boring, incremental reform of lightbulb efficiency standards gave rise to a hysterical popular backlash in the US. This resulted in "freedom of choice" lightbulb laws in conservative states and an ultimately successful twelve-year campaign to repeal federal standards. See e.g. Derysh 2019, or, for a broad overview, Wikipedia contributors 2020.

[14] Union of Concerned Scientists 2017; Davenport 2018; Irfan 2018.

[15] See e.g. Starostinetskaya 2019. Cultural constructions of masculinity can play a major role in steering democratic publics away from emissions-heavy (i.e. meat-centered) diets, as evinced by the extensive public debates that occurred in France in 2022 (Cohen 2022). So, too, can the romanticization of farming and farmers, as recent experiences in the Netherlands and Belgium suggest (Biesemans and Rossignol 2023; Coates 2023; Hayden 2023).

[16] On this possibility, see e.g. Dobson 1996; Holden 2002 (esp. ch. 3); Eckersley 2004 (esp. ch. 5).

[17] As suggested in Leopold 1989.

influencing normative assessments of legitimacy and are likely to only grow more salient as the climate crisis deepens.

Before moving on, three brief clarifications are necessary. First, I separate here questions about legitimacy from those concerning political obligation, and remain neutral about their relationship. Second, this chapter examines political legitimacy as a normative concept,[18] but with the understanding that the values underpinning it are determined by the perspectives, experiences, and assessments of specific human communities in specific historical moments, and thus are mutable in ways that many moral theories resist.[19] Finally, I use "authoritarian" in a fairly generic and expansive sense throughout to refer to political arrangements or modes of governance that are illiberal (i.e. rights- and freedom-constraining), undemocratic, and characterized by a concentration of executive power.

3.2 Two levels of political legitimacy

Our first task is to clarify what exactly is meant by political legitimacy. In this section, I examine a range of standard conceptions. My aim here is not to give an exhaustive overview of the (considerable) literature on legitimacy, nor to endorse or critique any particular theory, but rather to establish a useful conceptual distinction—that between FL and CL—in order to facilitate deeper analysis of authoritarian emergency power.

3.2.1 Foundational legitimacy

As noted earlier, FL pertains to a government's ability to ensure the safety and security of its citizens. One key threat here is that posed by political catastrophe, which, as discussed in Chapters 1 and 2, refers to situations in which material scarcity becomes so extreme that meeting one's own needs necessitates denying others the ability to do the same; under such zero-sum conditions, the ordinary principles and rules of justice break down, making peaceable social cooperation and political stability difficult, if not impossible, to maintain. Preventing political catastrophe—and, more broadly, ensuring basic safety and security—is bound up with a range of political capacities and actions, including, among other things, sustaining sufficient and continuous access to essential goods (particularly food, water, and shelter), providing immediate and effective disaster relief, and combatting invading forces or unjustified uprisings. If a government cannot fulfill these basic security functions, it is not legitimate, if it is even a government at all.

[18] As opposed to a "descriptive" one, of the sort commonly associated with Weber (2009 [1991]).
[19] Here I follow Williams 2005: 10.

This security-centric view is most naturally associated with Thomas Hobbes, who famously held that a legitimate government is one that is able to end the war of all against all and establish the conditions necessary for lasting peace and fruitful social cooperation.[20] These are such great goods that we should be willing (Hobbes thinks) to accept any power capable of providing them. In fact, on his view, the greater the power, the better, as this entails a more stable and enduring provisioning; hence *Leviathan*'s defense of absolute sovereignty. Hobbes concedes that unchecked power may result in a range of evils, like persecution of the innocent; yet he maintains that such risks are acceptable to the extent they are necessary for ensuring each person's overriding interest in his or her own preservation.[21] (For Hobbes, our interest in self-preservation is so fundamental that it provides the limit of political obedience: we should submit to political authority only up to the point that doing so will bring about our personal demise.[22]) Thus, so long as a given government has the *power* to keep the peace and protect its citizens—which, for Hobbes, constitutes "the very essence of government"[23]—its use of that power is, *ipso facto*, legitimate.

There are, of course, serious issues with Hobbes's view, most of which have been examined extensively before.[24] But a relevant problem for our purposes is that it is not immediately clear *who* is able to determine whether the sovereign is (capable of) protecting its citizens and therefore satisfying FL. Moreover, Hobbes affords little recourse for those subjected to a government that fails in this regard: he just repeatedly claims that only a greater power can oversee and check the sovereign,[25] in many instances suggesting that this greater power may be God alone.[26]

Even if we could find satisfying solutions to these problems, we should still reject Hobbes's more general view that ensuring the safety and security of those subject to a political power is *enough* for establishing the legitimacy of that power—at least given normal, reasonably favorable conditions. Bernard Williams makes this point. Like Hobbes, he believes that "the securing of order, protection, safety, trust, and the conditions of cooperation" comprises "the 'first' question of politics." Yet "first" in this context simply means that solving it is "the condition of solving" other important political questions—i.e. those at the level of CL. Thus, for Williams, in contrast to Hobbes, ensuring basic safety is a "*necessary* condition of legitimacy,"

[20] See e.g. Hobbes 1994: ch. 30.1.

[21] "And though of so unlimited a power men may fancy many evil consequences, yet the consequences of the want of it, which is perpetual war of every man against his neighbour, are much worse" (Hobbes 1994: ch. 20.18; see also chs 21.11–16, 14.29–30, 20.10, 20.18, 31.5).

[22] See e.g. Hobbes 1994: ch. 21.21.

[23] Hobbes 1994: ch. 21.17.

[24] The literature on Hobbes is too voluminous to cite comprehensively. For a helpful and highly accessible treatment, see Klosko 2013: ch. 2; for a compelling and concise critique, see Korsgaard 1997: 25–30.

[25] Hobbes 1994: ch. 20.18.

[26] Hobbes 1994: ch. 30.1.

but it is not (typically) "*sufficient*."[27] Rather, he argues, legitimacy also requires that power be exercised in broadly "acceptable ways."

The idea that in—and especially after—attending to order and security, broader questions about acceptability arise, is another way of indicating that legitimacy has a two-tiered dynamic. In this sense, the question of what counts as an acceptable use of power is, at bottom, one about which factors are relevant to CL. I examine the concept of CL in section 3.2.2. We should note here, though, as Williams is quick to do, that what counts as "acceptable" will vary considerably across cultural and generational lines, and especially in light of different circumstances. Thus, in our contemporary world, democracy and liberal rights are widely regarded as essential facets of acceptability, such that most, if not all, illiberal or non-democratic regimes are considered illegitimate.[28] Yet, in premodern times, other factors were considered central to acceptability, like the religious identification or familial lineage of the rulers.

This raises the obvious question: if contemporary conditions were to change, would rights and democracy remain central to legitimacy? While Williams never addresses this question directly, he does suggest in one passage that, in Hobbes's historical context, ensuring safety (by concluding the "war of all against all") may well have been sufficient for legitimacy. This observation raises further questions, however. What was it about Hobbes's time that made FL enough for legitimacy? Could similar conditions arise again today?[29] Moreover, how should a government proceed if prevailing standards of acceptability jeopardize or directly undermine its ability to ensure basic safety?[30] What if, for instance, a majority of citizens decide democratically *not* to redress an emerging existential threat? Should this seem implausible, consider how many thousands of people organized, protested, and voted so as to frustrate or defeat even the most minimal, low-cost public-health policies during the COVID-19 pandemic—e.g. requirements to wear a face mask or maintain a safe distance in social settings.[31]

We should note finally that Hobbes's and Williams's basic claim—that a government's ability to protect its citizens is an essential part of what makes it legitimate—is underspecified. It is reasonable to think that what matters for legitimacy is not just a government's ability to protect citizens *now*, but also its ability to continue doing so into the future.[32] This becomes clearer when we consider how much people are often willing to sacrifice (their time, their property, their

[27] Emphasis added; Williams 2005: 3.

[28] Williams 2005: 7ff., 15.

[29] I take up this question in section 3.4.1.

[30] I attend to this in section 3.3, though of course much the same question comprises the focus of Chapter 2.

[31] While this tendency has been most visible in the US, the same has transpired in Germany, Brazil, India, and many places besides.

[32] To be clear, none of the thinkers I have been discussing explicitly deny the relevance of this forward-looking dimension of FL, but all fail to examine it.

lives, etc.) to ensure that their society or nation will persist (and ideally flourish) after their own deaths.[33] This suggests that FL is intimately related to a government's long-term capacities and potential to endure, which centrally involves its ability to satisfy the precautionary duties laid out in Chapter 2. Put another way, even if a state is characterized by peace and material sufficiency in the present, should it be sufficiently vulnerable to future threats—such that at any point it could plausibly no longer be able respond to an emergency, maintain order, and protect its citizens—then that state may already be thought to lack FL. As Hobbes, Williams, and others emphasize, legitimacy arises as a function of power; a powerless government therefore cannot be legitimate. Present conditions and security capacities, then, are not enough for establishing FL. It must also be clear that the government is committed, at minimum, to not actively undermining future security; and, more substantively, to taking any necessary (preventative or redressive) action against critical threats, if and when they should arise.

To summarize, then, a government possesses FL if it can and does protect its citizens, and has (or is at least believed to have) the power and will to continue doing so into the future. Yet this is not (always) sufficient for political legitimacy. Other factors, related to what makes the use of political power acceptable, (typically) also matter. Important questions remain, however, about exceptional circumstances—i.e. conditions under which FL may alone be sufficient. In sections 3.2.2 and 3.3, I approach these questions by way of examining contemporary standards of acceptability and their relationship to FL.

3.2.2 Contingent legitimacy

According to theories of what I am calling CL, a government is legitimate if it exercises power in acceptable ways.[34] What is capable of rendering power acceptable is a matter of extensive debate, which is evinced by the fact that much of the literature on legitimacy centers on this question. In that work, one can find a broad array of factors defended, some of which are procedural, others substantive.[35]

Concerning procedural factors, a common view is that democracy is essential for legitimacy. The precise reasons for this vary among thinkers. Some, like Thomas Christiano, hold that democracies provide the most fair and effective means for resolving evaluative disagreements among equals.[36] Allen Buchanan,

[33] On the ethical significance of belief in a future after one's own death, see Scheffler 2016.

[34] There are many such theories, an exhaustive account of which I cannot provide here. For some emblematic views, however, see e.g. Nagel 1987; Rawls 1999; Buchanan 2002; Ripstein 2004. But cf. Simmons 1999, who distinguishes legitimacy from the "justification of power."

[35] Alternatively, following Scharpf (1999), the relevant distinction could be described as that between input- and output-oriented bases of legitimation. I thank an anonymous reviewer for this point.

[36] See e.g. Christiano 1996: 35. See also Gaus 1997; Waldron 1999; Christiano 2008. Cf. Estlund 2008.

on the other hand, argues that wielding political power over equals can only ever be justified if each citizen has an "equal say" in determining who will command that power, and democracy alone can provide this.[37] On both of these accounts, however, and many others besides, the basic conclusion is the same: legitimacy requires democratic processes and institutions.

Many also associate legitimacy with one or more substantive factors. One enduringly popular view centers on consent, taking this as necessary for squaring the imposition of coercive power with individual freedom and the moral equality of persons.[38] This view is often criticized, however, on the ground that very few people ever actually consent to a government in the explicit and content-independent sense thought necessary for establishing political legitimacy. This difficulty has led sympathetic scholars to either adopt a mediated notion of consent—e.g. tacit or hypothetical consent[39]—or else insist that consent matters for grounding a government's "claim-right to obedience" but is irrelevant to the broader question of legitimacy.[40] Many others, however, maintain that, given consent theory's problems, other factors are more relevant to legitimacy. Perhaps the most compelling alternative holds that a government is legitimate if it wields its power (a) so as to protect the most basic rights of those who are subject to it, and (b) in such a way that expresses respect for those rights.[41] This "basic- rights view" is understood to imply strict constitutional limits on governmental power. And it is popular among liberal thinkers because it avoids the obvious pitfalls of consent theory without resorting to consequentialism.

There are of course many other procedural and substantive factors associated with legitimacy, including, for example, equal representation, provision of core public benefits, adherence to social-justice norms, and distributive fairness. This very cursory survey, however, should suffice for examining three general issues facing all CL theories. First, as noted earlier, there is widespread disagreement about which factors are (most) relevant to legitimacy. Moreover, most people—including most philosophers—maintain that legitimacy requires multiple factors (e.g. democracy, fairness, *and* guaranteed freedoms). This raises the problem of determining which matter(s) most, particularly in instances of conflict.[42] Should it be the case, for instance, that fair and open democratic processes yield results that undermine individual rights—as when majorities vote to limit free expressions of

[37] Buchanan 2002.

[38] This view is widely held among early social-contract theorists (e.g. Hobbes, Locke, and Rousseau), and more recently among Lockean thinkers like Nozick (1974) and Simmons (2001).

[39] See, e.g., Hill 2001; Peter 2017. Estlund's 'normative consent' model also falls into this category (2008: 117*ff*). Cf. Dworkin 1975: esp. 17-21; Wellman 1996; Pateman 1988.

[40] Copp 1999; Buchanan 2002. See also Raz 1995: 356f. for further discussion.

[41] Buchanan 2002: 703.

[42] Consider, for instance, the debate among "minimalist" theories of legitimacy, which hold that guarantees of basic human rights are sufficient for legitimacy, and more expansive accounts, which hold that democratic rights or processes are also necessary (Meckled-Garcia 2014).

faith[43]—reasonable people might disagree about whether such a decision is legitimate. We can refer to this as the *problem of pluralism*; it suggests that any firm agreement on what matters (most) for CL will remain elusive, or at least always open to contestation and revision.

The pluralism and so contestability of CL is likely a function of its historically determined character: bases of legitimation at this level come into existence slowly, often only partially supplanting others. For instance, before Hugo Grotius, Samuel Pufendorf, and the broader social-contract tradition emerged, there was little concern over whether states enjoyed the consent of those they governed; other standards were thought to ground legitimacy.[44] Robert Filmer, for one, argued that what makes a government legitimate is the ability of leaders to trace their lineages back to Adam.[45] While this view may appear ridiculous, we must remember that lineage and religious affiliation were among the most salient CL factors until quite recently, and undoubtedly continue to shape perceptions of legitimacy today. Consider, for instance, the right-wing obsession with former US President Barack Obama's religious affiliation:[46] the mere possibility that he was Muslim was thought by many to disqualify him from the office. Although this may seem a fringe preoccupation, opinion polls routinely show that religious identification matters greatly to many US citizens. (According to one survey, nearly one-third of voters believe that atheists should be legally prohibited from becoming president.[47]) The slow, historical process of value-change means that CL theories also face what could be called the *problem of partial displacement*. In other words, when new legitimation factors emerge, earlier ones may not entirely disappear but only become less salient, at least for sizable portions of the citizenry.[48] This further compounds the problem of pluralism noted earlier.

A third and final issue concerns the fact that even widely shared and seemingly stable CL factors are routinely relaxed or abandoned during emergencies, often without calling into question the basic legitimacy of the government. We can refer to this as the *problem of exceptional circumstances*, examples of which are manifold.[49] A familiar case concerns the authoritarian measures Abraham Lincoln took during the US Civil War. As George Kateb describes: "Lincoln destroyed the revered Constitution temporarily to save it." This "destruction" involved suspending or failing to protect some of the most important rights that document

[43] Consider, for example, Switzerland's 2009 "minaret ban," approved by popular referendum, or France's 2010 "burqa ban" (*loi* n° 2010–1192).

[44] Peter 2017.

[45] This idea of legitimacy-by-birthright survives, to some extent, in the way many cultures and religious groups think about paternity: i.e. a child is "illegitimate" if it is not born to two wedded parents.

[46] The racist conspiracies surrounding his birthplace, which never denied that he was raised and fully educated in the US, offer another case in point.

[47] Cox 2017; Freedland 2017.

[48] For an argument along these lines, see Mittiga 2019b: esp. 188ff.

[49] See e.g. Rossiter 1948; Lazar 2009.

enumerates—many of which, like habeas corpus, were then, as they are now, considered fundaments of political legitimacy. The COVID-19 pandemic represents another, more contemporary example: in liberal-democratic states across the world, elections have been delayed, domestic and international movement severely restricted, associations and meetings limited or banned, commerce constrained, private factories repurposed by fiat, and in some cases free speech curtailed to prevent the spread of disinformation. Such authoritarian impositions are common during wars, economic depressions, and wide-scale natural disasters. Of course, not all emergencies are equal—or even *real*; as such, these powers have a high potential for abuse.[50] Moreover, even when they do not function as a vehicle of corruption, the use of authoritarian powers may still precipitate or exacerbate political instability.[51] Nevertheless, the fact that even liberal-democratic governments consistently resort to such measures in moments of great exigency suggests that the value of CL factors greatly depends upon the prevailing circumstances.

Understanding the problems of pluralism, partial displacement, and exceptional circumstances makes clear why CL factors are rightly considered contingent: views about what makes the imposition of coercive political power acceptable vary tremendously across generations, cultures, states, and circumstances. Unless we adopt a Whiggish view of history, believing that, while only being "discovered" recently, contemporary CL standards are nonetheless universal constants, then we must conclude that what is often regarded as *axiomatically* true today—that, say, only rights-respecting/consent-based/democratic governments are legitimate—is actually only *contingently* so: i.e. true for those living in secular, relatively affluent, Western states. Quite different standards of legitimacy are and have been held, with equally great force, in different cultures and epochs, and still others are likely to emerge in the future.

3.3 The relationship between foundational and contingent legitimacy

We must now consider the relationship between FL and CL. This relationship has not been well explored in the literature on legitimacy, perhaps because under reasonably favorable conditions—which are common today in the places where much of the world's philosophy is produced—serious tensions between CL and FL tend not to arise. Rather, under such favorable conditions, the imperative of maintaining safety can be satisfied without violating CL factors. This, again, appears to be Williams's point in claiming that FL represents only "the first political question": i.e. that once basic safety and security are assured, concerns about justice, rights,

[50] As briefly discussed in Chapter 2, section 2.6.4.
[51] See e.g. Rossiter 1948; Fassin and Pandolfi 2010.

and democracy take center stage and legitimacy assumes a more complex, moral shape. In short, FL operates (often unacknowledged) in the background, being more invisible the more firmly security is established.

The possibility of conflict becomes obvious in crisis moments, however, during which the ability of a government to protect its citizens is no longer assured. Following the literature, we may refer to such moments as "states of exception."[52] A state of exception (hereafter, SOE) is precipitated by an emergency (or credible threat thereof) of sufficiently great magnitude that prevailing political institutions, processes, norms, etc. either impede the swift action needed to preserve/restore normal conditions, or else simply break down.[53] Faced with an SOE, political leaders have a terrible choice: either (a) relax or suspend any laws, processes, norms, etc. that hinder action—many of which may be central to CL—in order to respond expediently and effectively to the threat; or (b) proceed with normal constraints intact and risk danger, destruction, and collapse.[54]

The examples provided above, concerning political responses to the US Civil War and the COVID-19 pandemic, illustrate this dilemma. There are of course countless others, tracing back to the dictatorships of ancient Rome.[55] My aim here, however, is not to enumerate types of emergency powers, nor to consider whether their use was or could have been justified in any particular case. Rather, I am interested in examining the tension that arises from the fact that, during SOEs, the institutional and normative structures that are meant to safeguard individual rights, democratic governance, and so on, may actually serve to hinder necessary action, and to consider what this means for legitimacy.

My essential claim is that when such conflicts arise, political legitimacy will depend more (or even exclusively) on whether security needs are met than on whether CL factors are satisfied. Establishing this requires us to pursue two basic questions: (1) what, exactly, justifies prioritizing FL over CL when conflicts between them occur?; and (2) who decides when and to what extent such trade-offs are permissible?

Concerning the first question, we must recall the basic difference between CL and FL. The factors associated with CL condition the use of political power by specifying, for instance, what can or cannot be done or sacrificed, how decisions should be made, and who counts (and for how much). The answers to these questions often appear to us as moral universals; yet, in practice, they are the products of long and contentious historical processes. FL, on the other hand, does not vary between societies, generations, or circumstances. Ensuring safety and security is

[52] See e.g. Schmitt 1985; Agamben 2005; but cf. Lazar 2009 (esp. ch. 2) and Scheuerman 2006, who complicate use of this term. Another possibility, following Walzer (2006: ch. 16), would be to refer to them as "supreme emergencies."

[53] Schmitt 1985: ch. 1; Agamben 2005.

[54] As Rossiter (1948: 290) aptly puts it: "Into whatever forbidden fields of freedom the necessities of crisis may force the leaders of a constitutional government to go, go they must or permit the destruction of the state and its freedom."

[55] See e.g. Rossiter 1948; Machiavelli 1994, *Discourses on Livy*, Bk. I, ch. 34.

always the *primary*—though, in good states, under reasonably favorable conditions, not the *exclusive*—end of political power. Aristotle expresses something like this in insisting that the point of political society is to furnish the resources needed not just to live, but to live *well*.[56] Crudely put, FL is about living, CL about living well. And it is of course impossible to live well without living: there can be no democracy of desolation, no fair social cooperation in conditions of extreme scarcity, no real rights when political stability is maintainable only through raw assertions of coercive power (if it can be maintained at all). In this sense, FL is *necessarily* prior to CL, and must be regarded as such in moments when trade-offs become a necessary part of the political calculus.[57]

One might object here that a government's ability to keep us alive, however wretchedly, is *not* enough for legitimacy, even in thrall to a genuine SOE. Consider Patrick Henry's compelling plea to the Second Virginia Convention in 1775:

> Is life so dear or peace so sweet as to be purchased at the price of chains and slavery? Forbid it, Almighty God! I know not what course others may take; but as for me, give me liberty or give me death![58]

Clearly, there are some values for which many people, like Henry, would willingly risk grave harm to protect or promote. But this fact is not in tension with my argument. Henry is criticizing the idea of authoritarian power being used in normal (non-SOE) conditions to maintain an artificial peace. My argument, however, is that relaxing or suspending CL standards is justifiable only when—and to the extent that—doing so is necessary to address serious and credible threats to citizens' safety.

Moreover, as both Hobbes and Williams stress, a legitimate government should not be the source of the security threat—the very "reign of terror"—its own power was meant to conclude.[59] This would likely be the case, as it was for Henry, if a government abridges rights or circumvents democratic processes for reasons other than responding to a threat at the level of FL. The Chilean uprising ("*estallido social*") of 2019 offers a contemporary example of this point. In October of that year, the government deployed militarized police forces to curtail peaceful student protests against a sudden metro fare increase. But the police violence and the democratic suppression it entailed only served to catalyze additional, much larger protest actions. While most of these were peaceful, some, especially during the first two days, did involve destruction of property and violence (particularly against police officers). Rather than attempting to de-escalate the situation

[56] See e.g. Aristotle 1998, Bk. 1, ch. 2, esp. 1252b25–35.

[57] The essential idea here is captured well by Bertolt Brecht's famous remark: "grub first, then ethics." I thank an anonymous reviewer for this point.

[58] Quoted in Wirt 1817.

[59] Williams 2005: 4.

through deliberation, ameliorative social policies, or even modest public apologies, the Piñera government instead responded by declaring martial law, greatly increasing the presence of armed police forces, and implementing extreme curfews. These were authoritarian measures, which—the government claimed—were justified in light of the security threat posed by the massive public demonstrations. In the framework I have been defending, however, such impositions could not have been considered legitimate, given that, again, the government's own actions were and continued to be the real and direct source of the security threat, and because other rights- and democracy-respecting (i.e. non-authoritarian) solutions remained available. What ensued was not a security threat for the people, so much as for (some of) the government. And to the extent ordinary people were at risk, non-authoritarian means could have been used to defuse tensions and resolve the underlying danger.

This suggests another important point. I am not arguing that only unchecked authoritarian power is appropriate during a SOE. Sensible limits may be observed, which may effectively operate as temporary alternative bases of CL. In the Roman Republic, for instance, dictatorial powers were subject to regular assessment and renewal by the Senate, and so limited in time and scope.[60] Similarly, most contemporary political societies require that a transparent standard be set for determining when a serious threat arises (and when it has passed), before ordinary CL factors can be justifiably overridden.[61] These measures may (at least sometimes) prevent "emergencies" from being declared against amorphous entities, like terrorism, the threat of which has sustained a continuous "SOE" in the US since 2001.[62] The literatures on just war and political violence suggest other plausible constraints.[63] We might, for instance, stipulate that authoritarian interventions must be *necessary*—i.e. that no other viable options remain for responding to the crisis at the speed or scale required. Also relevant would be a version of the *success* criterion, which holds that the exercise of such powers must have a reasonable chance of achieving their goal (i.e. overcoming the crisis). Another candidate would be something like the *proportionality* constraint, in this case specifying that authoritarian power should be exercised only to the extent necessary for, and in the domains relevant to, responding to the crisis.[64] These points notwithstanding, it is important to note that limits on emergency powers are *not* coincident with ordinary CL factors. Moreover, in many cases, even legally constrained emergency powers may still be fully compatible with authoritarian governance.[65]

[60] See e.g. Rossiter 1948; Lazar 2009: esp. ch. 5.

[61] Scheuerman 2006: 76*ff*; Lazar 2009: 133.

[62] See e.g. Korte 2017; Lu 2019.

[63] The following three standards follow Pasternak's (2018) innovative adaptations of the same.

[64] Still other standards might be defended, for instance concerning transparency and accountability. Here, the literature on states of emergency is instructive (e.g. Scheuerman 2006; Lazar 2009).

[65] As Lazar (2009: esp. 141–7) shows.

This suggests the significance of the second question above, about who decides when a state is facing a SOE and which trade-offs are necessary for addressing it. The best-known answer to this question is Carl Schmitt's, who defined sovereignty precisely in terms of the capacity to "decide on the exception."[66] This stands behind his critique of liberal democracy, which he believes lacks recourse to a single agent capable of declaring a SOE and deciding which, if any, CL factors must be relaxed or suspended in order to preserve or restore safety. This is a serious defect, as it means that such governments may be unable to respond quickly and effectively to emergencies when they arise.

While this concern may be plausible in relation to an idealized liberal-democratic state, in politics, just as in nature, there are no pure forms. Virtually every government today has explicit provisions for emergency rule.[67] For instance, the US National Emergencies Act affords the president a range of powers, including the ability to declare martial law, seize land, suspend the internet, and indefinitely detain suspects without criminal charges.[68] Similar measures exist even in liberal-democratic states with a deep cultural-historical wariness of authoritarian power. This is true in Germany, for instance, which long resisted formally promulgating such powers, given that similar constitutional provisions—the now infamous Article 48—provided a fully legal route for the Weimar government's descent into fascist totalitarianism.[69]

Many states only ambiguously define the full scope of emergency powers an executive can wield, as well as the conditions that can trigger their use. In some sense, this is to be expected: it is difficult to specify in advance what shape an emergency will take and how best to respond to it. As Schmitt stresses, the SOE "defies general codification."[70] We can isolate some common features, however. In SOEs, legislative and judicial powers, along with ordinary constitutional checks and balances and individual rights, are often (to varying degrees) relaxed or suspended, while a unitary executive is greatly empowered (though typically within some time limit or subject to some more or less transparent oversight agency). In this way, the promulgation and use of emergency powers appears to blur the line between constitutional forms. Hence Agamben's perceptive observation that the SOE "appears as a threshold of indeterminacy between democracy and absolutism."[71]

[66] Schmitt 1985: 5.

[67] Rossiter 1948; Agamben 2005: ch. 1; Kreuder-Sonnen 2019.

[68] Brennan Center for Justice 2018; Goitein 2019.

[69] See Herf 2019. Germany only relented when the Allied powers made constitutional provision of emergency powers a condition of its full independence, on the ground that Allied troops stationed there would be vulnerable unless the German government possessed the legal means to protect them in exceptional circumstances. This external pressure, in addition to the internal pressure brought on by the massive protests of 1968, eventually led to the adoption of the *Notstandsgesetze* (the Germany Emergency Act).

[70] Schmitt 1985: 13.

[71] Agamben 2005: 3.

Critically, this suggests that liberal democracies do *not* have a monopoly on political legitimacy. Quite the opposite. The significant extent to which liberal-democratic governments avail themselves of consolidated, extralegal power to address crisis situations is a tacit acknowledgment that maintaining legitimacy (in the foundational sense) requires, at times, some infusion of authoritarianism. This, of course, should not cause us to dismiss the differences between constitutional forms in normal conditions, or to overlook the fact that governments can be *more* or *less* legitimate than others, even during SOEs. It does suggest, however, that a capacity for authoritarian governance, as well as a means for expeditiously adopting it in dire circumstances, constitute essential components of political legitimacy.

3.4 Climate change and legitimacy

So far, I have argued that FL and CL ordinarily exist in harmony with one another. During, or in anticipation of, SOEs, however, conflicts between FL and CL can and often do arise—particularly when factors associated with the latter (like democratic processes or individual rights) limit a government's ability to preserve or restore basic safety. In these situations, FL is necessarily prior to CL, as CL factors cannot be sustained in conditions of great insecurity or material scarcity. Put another way, without FL, there can be no CL, at least for long.

Prioritizing FL over CL in crisis moments may entail embracing authoritarian power. In this sense, authoritarianism is *not* antithetical to political legitimacy; rather, legitimacy seems to require it, at least in exceptional circumstances. Again, this claim should not be taken as a denial of the potential dangers of authoritarianism, the vast potential for abuse it presents, or the extent to which emergencies can be fabricated to facilitate the consolidation of political power. Authoritarian governance is a blunt instrument, only to be wielded in times of great exigency.

3.4.1 Climate emergency

Typically, the best means we have for avoiding authoritarian impositions is preventing, as far as we are able, the kinds of situations that make recourse to them necessary. Unfortunately, after decades of inaction, this may no longer be possible with respect to climate change,[72] which today confronts us as an emergency in two senses of the word.

[72] While I focus here exclusively on climate change, Jocelyn Stacey (2018) has persuasively argued that all environmental issues should be approached through the "perspective of an ongoing emergency." This is not (just) because many contemporary environmental issues constitute "emergencies in a conventional sense," but because, like emergencies, "the complex, adaptive nature of all environmental

3.4.1.1 Emergency of effects

First, and most obviously, climate change represents an *emergency of effects*. As recent IPCC reports show, in many regions and states, climate change is already causing massive displacement, agricultural loss, famine and drought, extreme weather, and novel health crises. Without sufficient action, by the end of the century these and other "extreme events" will become consistent and crippling global phenomena.[73] Indeed, according to a (likely conservative[74]) report from the World Bank, "we're on track for a 4°C warmer world [by 2100] marked by extreme heat-waves, declining global food stocks, loss of ecosystems and biodiversity, and life-threatening sea level rise."[75] These conclusions have been supported by a number of more recent studies and reports, including the important 2019 IPCC Special Report on *Climate Change and Land*.[76] The reasons for this are manifold, and explained at length in Chapter 1 and the Appendix; we can briefly consider a few salient points here, however.

For one, climate change is causing increasingly serious and frequent shocks to global food and water supplies, which threaten new and extreme forms of scarcity, and thus political catastrophe, on a massive scale.[77] With the warming we have already "locked in," for instance, approximately one-third of the glaciers in the Himalayan region, which today provide drinking water for nearly 2 billion people, are likely to disappear over the next several decades.[78] Changes in precipitation, along with increased temperatures, are similarly affecting access to fresh water and disrupting agriculture in other parts of the world. According to one study, without dramatic reductions in emissions, by 2080 "[n]one of [the] places, which today supply much of the world's food, will be reliable sources of any."[79] And of course this is

issues precludes reliably predicting which ... contain the possibility of catastrophe" as well as our ability to determine "in advance how we ought to respond once the possibility of a catastrophe becomes apparent" Stacey (2018: 15–16). These epistemic challenges suggest that the emergency governance framework may be enduringly relevant to environmental politics.

[73] IPCC 2021b.

[74] Brysse, et al. 2013; Oreskes, et al. 2019.

[75] World Bank 2012a: n.p, referring to World Bank 2012b. Of course, since this report, many states have committed themselves to taking greater action—most significantly, in accordance with their "nationally determined contributions" (NDCs) to the 2015 Paris Agreement (many of which were subsequently expanded during COP26, the 2021 UN Climate Change Conference of the Parties). Yet, as the latest IPCC reports (AR6) and other studies have shown, even assuming the (COP26-enchanced) NDCs were fully realized—which, given current action deficits, is by no means certain for most countries (IPCC 2018c: 15; Plumer and Popovich 2021)—they would still fall significantly below what is needed to limit warming within 1.5°C, or even 2°C (IPCC 2022b: 19; Liu and Raftery 2021). In fact, on all but the lowest emissions scenarios, the IPCC (2021a: 18) predicts median warming of between 2.7° and 4.4°C by 2100—which, even if *potentially* below the 4°C of warming the World Bank warned against, would nonetheless be enough to precipitate extreme and enduring scarcity (and thus political catastrophe) in many parts of the world.

[76] IPCC 2018a.

[77] For a more developed defense of this claim, see the Appendix and Chapter 2, especially section 2.6.1. See also Mehrabi 2020; Kornhuber, et al. 2023; Richards, et al. 2023.

[78] Wester, et al. 2019.

[79] Wallace-Wells 2017, citing Cook, et al. 2014. See also IPCC 2018a: esp. ch. 5 ("Food Security").

unfolding against a backdrop of explosive population growth, which, according to the UN and various experts, will require a *doubling* of food production by 2050.[80]

The extreme heatwaves associated with climate change, along with the pollution driving it, are already causing millions of premature deaths each year, and this will only become worse going forward.[81] This tragedy is compounded by the unequal distribution of climate effects globally. For instance, according to Nicholas Stern, "compared with a world without climate change," 2° to 3°C of warming by 2100 will cause the deaths of "an *additional* 165000–250000 children" each year in South Asia and sub-Saharan Africa alone.[82] This helps to explain why a Marshall Islands representative described the Paris Agreement's modest pledges as "genocidal,"[83] and why the UN Secretary General, António Guterres, called a recent IPCC report a "code red for humanity."[84]

These developments are likely to aggravate preexisting stressors such as poverty, environmental degradation, and socio-political tensions, and thereby trigger more frequent and aggressive resource conflicts.[85] According to one estimate, for every 0.5°C of warming, armed conflict is 10–20 percent more likely to occur.[86] For this reason, global security experts routinely refer to climate change as a "threat multiplier."[87]

3.4.1.2 Emergency of action

The time window for preventing these dreary outcomes is rapidly closing; in this sense, climate change also constitutes an *emergency of action*. According to the IPCC, to have a better-than 50 percent chance of preventing a >1.5°C rise in global temperatures—which is widely regarded as the upper limit of noncatastrophic warming—we must reduce global emissions approximately 45 percent by 2030, and achieve net-zero emissions (i.e. complete carbon neutrality) by 2050.[88] This would require an unprecedented mobilization of resources

[80] United Nations 2009; Tilman, et al. 2011; Ray, et al. 2013; but cf. van Dijk, et al. 2021. For further discussion, see the Appendix. Of course, doubling food production would almost certainly come at the cost of further deforestation, which already is nearing a dangerous crisis point. According to a recent peer-reviewed study in *Nature*, should "the actual rate of population growth and resource consumption, in particular forest consumption," continue unchanged, we have a *less than 10 percent chance* "in the most optimistic scenario" of surviving "an irreversible collapse of our civilization" within the next four decades (Bologna and Aquino 2020).

[81] Burnett, et al. 2018; Vohra, et al. 2021; Zhao, et al. 2021; and, perhaps most importantly, Bressler 2021.

[82] Emphasis added; Stern 2007: 55.

[83] Pashley 2015; Wallace-Wells 2018.

[84] United Nations 2021.

[85] For further discussion of this point, see Chapter 1.

[86] Burke, et al. 2015 and Carleton, et al. 2016 (both cited in Wallace-Wells 2017). In some cases, temperature increases alone are enough to spur aggression. According to one report, we could see as much as a 54 percent increase in armed conflict in sub-Saharan Africa by 2030 (Burke, et al. 2009).

[87] US Department of Defense 2014b. NATO likewise refers to climate change as a "threat multiplier" (Nuccitelli 2017). See also The White House 2022: 9.

[88] IPCC 2018b: 12; Watts 2018. This has been reaffirmed in IPCC 2021b: section B and IPCC 2022b: 21–25.

and political will—likely exceeding that undertaken during World War II[89]—particularly among the world's most heavily polluting (and typically most recalcitrant) states, like the US, China, Australia, Canada, Saudi Arabia, Brazil, and India. Moreover, the longer action is delayed—which is to say, the longer states shirk their precautionary duties[90]—the more rapid and extreme action will have to be later.[91]

In any case, the necessary policies—which would almost certainly involve enormous public investment, and so new taxes or sovereign debt, along with expansive restrictions on individual behaviors and consumption—are not likely to be popular, and may indeed inspire considerable resistance, both among the general public and negatively impacted industries. Nevertheless, preventing catastrophic climate change depends on governments' ability (and willingness) to swiftly promulgate, widely implement, and consistently enforce such policies, particularly during the next three decades.

3.4.1.3 Climate action and regime legitimacy

Given all of this, it is clear why, often at the behest of committed activists, "2,275 jurisdictions in 39 countries," including "18 national governments and the EU"—together covering over 1 billion people—have now declared a "climate emergency."[92] Yet, while such declarations are symbolically significant, they have rarely been paired with ambitious, or even adequate, action. Consider the most recent IPCC report's finding that, even if states were to fully implement their Paris Agreement pledges, average global temperature would still "likely" exceed 1.5°C before 2100, and avoiding 2°C of warming would become extremely difficult. In fact, IPCC projections in the same report show that—again, even if the pledges were fully implemented—states in 2050 would still be emitting more than six times the (roughly 10GtCO2eq) maximum limit compatible with having a greater than 50 percent chance of limiting warming to 1.5°C.[93] If far more extensive action is not undertaken and soon, many states face the prospect of intense near-term shocks, scarcity, and violence; while, in the long term, all are risking political catastrophe and collapse.

In this sense, it appears that climate change constitutes a full-scale legitimacy crisis, even for otherwise legitimate governments. For if we take the imperative of ensuring safety and security to be a *necessary* condition of legitimacy, as I have been arguing we should, then it follows that those governments that do not

[89] This argument has been made many times in the last few years, particularly in relation to "Green New Deal" proposals (see e.g. McKibben 2016).

[90] For a discussion of these duties and what fulfilling them entails, see Chapter 2.

[91] IPCC 2022b: section B.6, pp. 18ff.

[92] As of September 2022, included here are Canada, the UK, Ireland, France, Argentina, Spain, and Austria, among others, as well as 1,261 local or municipal governments (Climate Emergency Declaration 2022).

[93] See the graph on IPCC 2022b: 19. See also IPCC 2021b, esp. p. 18, and Liu and Raftery 2021.

undertake adequate climate action are, for that reason, *illegitimate*, regardless of their commitments to liberal-democratic values or other normative virtues. Note, however, that the converse is not (necessarily) true: so long as it is possible to maintain liberal-democratic institutions, norms, and principles in at least some domains of political life, undertaking adequate climate action will not be *sufficient* for legitimacy. Consequently, however admirable its commitment to climate action may be, a fully authoritarian regime (like China's) cannot, on that ground alone, be considered normatively legitimate. This asymmetry may dissolve, however, if, some years from now, the climate crisis becomes so severe that nothing short of completely abandoning liberal democracy would be capable of preventing (or forestalling) political collapse.[94]

3.4.2 Authoritarian climate governance

I have argued that responding to the security threat posed by the climate crisis may legitimate resorts to authoritarianism, both in managing the fallout from impending or unfolding climate catastrophes (i.e. the emergency of effects) and in ensuring that such events are more limited in number and scope in the future (i.e. the emergency of action). Assuming this is right, we should now consider what authoritarian climate governance might entail. Although it is not possible to answer this question fully here, we can consider a few possibilities.

For one, governments might impel citizens to make significant lifestyle changes. One pertinent example concerns curbing meat-heavy diets, common in the Global North, given the enormous carbon footprint of animal agriculture.[95] Under normal conditions, any attempt to change how people eat would be considered an unacceptably paternalistic affront to individual autonomy. Yet there is by now extensive evidence that it is likely impossible to avoid catastrophic climate change without drastic reductions in animal agriculture.[96] Again, if such restrictions are necessary for preserving FL, then they may be justifiably imposed, even if doing so cuts against the wishes of democratic publics or violates individual or group-based rights.[97] Similarly, governments might consider restricting or even prohibiting elective air travel, given the outsized and growing emissions of that sector, again even if this constitutes a limit on the exercise of free movement rights.[98]

[94] Versions of this haunting possibility are explored in several recent publications (e.g. Davis 2010; Mulgan 2014; Oreskes and Conway 2014; Battistoni 2018). Perhaps most notable is Mann and Wainwright 2018, which identifies a form of global climate authoritarianism, dubbed "Climate Maoism," as one of four possible future scenarios.

[95] See e.g. Ripple, et al. 2014, Herrero, et al. 2016; IPCC 2018a: sections 1.2.1, 1.3.3, 2.3, 5.5.2, 5.6.5; Xu, et al. 2021; Scarborough, et al. 2023.

[96] Clark, et al. 2020; Harwatt, et al. 2020; Ripple, et al. 2022.

[97] For an argument along these lines, see Wood, et al. 2023.

[98] Higham, et al. 2019.

Policies aimed at limiting reproduction may also become necessary, particularly in high-emissions countries like the US.[99]

We may also imagine a censorship regime that prevents the proliferation of climate denialism or disinformation in public media. This may well conflict with standard conceptions of freedom of expression or of the press. Again, however, to the extent that those freedoms have been exercised in ways that have undermined (and continue to undermine) effective climate action,[100] such censorship may be warranted.[101] As Nomi Lazar argues, rights may "be derogated when other interests of critical importance are taken into account." This is especially clear, she thinks,

> for freedom of expression, where we habitually make restrictions on the basis of other values: no matter how free our speech, it is never so free that we can shout "Fire!" in a theater, thereby endangering the lives of patrons. Free speech is important, but so is the safeguarding of life.[102]

Likewise, responding effectively to climate change may require relaxing property rights in order to nationalize, shutter, or repurpose certain companies, particularly in the energy and agriculture sectors, so as to ensure that the transition away from carbon-intensive production happens rapidly, fairly, and at limited cost.[103]

Governments might also limit certain democratic institutions and processes to the extent these bear on the promulgation or implementation of essential climate policy. This could involve imposing a climate "litmus test" on those who seek public office, disqualifying anyone who has significant (relational or financial) ties to climate-harming industries or a history of climate denialism. More strongly, governments may establish institutions capable of reviewing and overturning previous democratic decisions (expressed, for example, in popular referenda or plebiscites) against, say, the implementation of carbon taxes, the shuttering of coal factories, or other necessary climate policies.

This, again, is not exhaustive of what authoritarian environmentalism may entail. But the key point here is that if, and to the extent that, ordinary CL factors

[99] Murtaugh and Schlax 2009; Wynes and Nicholas 2017.

[100] See Oreskes and Conway 2011.

[101] There is a direct analogy here to public-health disinformation campaigns. In reflecting over misinformation propagated on social media during the COVID-19 pandemic, for instance, Jeffery Howard (2021: 168) argues that "the moral right to freedom of speech, like all our moral rights, is limited by a moral duty we have not to impose undue risks of harm on others. When [social media] users post harmful misinformation they violate this duty, even if unintentionally. Accordingly, those who endanger others by posting misinformation are not wronged when their posts are removed."

[102] Lazar 2009: 91.

[103] Actions like this appear already to be occurring. Consider the Netherlands Supreme Court's May 2021 ruling that, by continuing to emit greenhouse gases in awareness of the harms involved, the oil company Royal Dutch Shell was in violation of Articles 2 and 8 of the European Convention on Human Rights (which guarantee the right to life and to family, respectively) (Boffey 2021). Despite Shell's protests, the company is now required to reduce its emissions by 45 percent by 2030, relative to 2019 levels.

inhibit urgently needed climate action, they may, on my account, be justifiably relaxed or suspended until the (credible threat of) emergency has passed. This is perhaps an unsettling conclusion, but at this late stage there may (soon) simply be no other means of ensuring safety and so satisfying the basic demand of FL.

3.5 Emerging bases of legitimacy

Responding to concerns about safety and security is not the only way in which climate change may affect political legitimacy. In addition to prompting temporary and local suspensions of prevailing liberal-democratic CL standards, the climate crisis may also give rise to *new* standards of legitimacy. Although I can only sketch this possibility here, historical experience suggests that grave crises can be generative in precisely this way.[104] Consider, for instance, the way in which the provision of basic economic rights—e.g. to healthcare, unemployment insurance, senior pensions—came to be associated with political legitimacy following the Great Depression. Certainly, not every state provides for these rights, or even tries to; but it is undeniable that many people—particularly those in advanced industrial societies—regard such economic safeguards as part of what makes their governments legitimate. In other words, states which (unnecessarily) perform poorly in this respect are widely considered *less* legitimate than states that can and do sustain a more generous suite of economic rights.

It is difficult to predict which new standards might arise through confrontation with climate change, but we can consider three possibilities here—all of which, I believe, are already influencing perceptions of legitimacy.[105] The first, which we can call the *stewardship standard*, holds that a government that fails to adequately represent and attend to the interests of future generations (and perhaps of nonhuman animals and ecosystems) is, for that reason, illegitimate (or less legitimate). A commitment to stewardship—and the underlying sense of obligation to those with no voice who are or will be affected by the climate crisis—helps to explain why activist and scholarly critiques of climate change so often focus on the dangers it poses to those outside of the contemporary human community. (Consider

[104] Rossiter (1948: 13, 295) claims that such "alterations" are inevitable after any large-scale crisis; however, he takes the (reasonably) pessimistic view that they tend to be "unfavorable." Here, I am interested in potentially *positive* changes that may arise, although I do not mean to suggest that such changes are inevitable or that significant negative developments will not (also) occur. As Rebecca Solnit (2016: 13) observes: "Inside the word 'emergency' is emerge. From an emergency, new things come forth. The old certainties are crumbling fast, but danger and possibilities are sisters."

[105] There is an interesting parallel between the emerging standards of political legitimacy I describe here and certain green republican ideals, like those exhorted by Barry and Smith (2008). This suggests that the future need not be considered purely in terms of maintaining prevailing forms of liberalism or succumbing to authoritarianism. Rather, to the extent that more ecocentric CL standards take hold, different political forms—including green republicanism or even something akin to Mann and Wainwright's (2018) post-capitalist "Climate X" scenario—may emerge.

Extinction Rebellion's "Declaration of Rebellion": "we declare it our duty to act on behalf of the security and well-being of our children, our communities and the future of the planet itself [...] We act on behalf of life."[106]) In this sense, the stewardship standard is also an important driver of demands for new political institutions capable of affording representation to ecosystems, nonhuman animals, and future people.[107] Clear examples of this are the recently constituted official positions in Wales (the "Future Generations Commissioner") and Sweden (informally, the "Minister for the Future"), both of which are tasked with ensuring that the interests of future citizens are taken into account in contemporary political institutions and forums.[108] The US has also recently designated, for the first time, "a diplomat for plants and animals" in the form of a special envoy for biodiversity attached to the State Department.[109]

Another potential base of CL is the *epistocracy standard*, which holds that only those governments that operate in accordance with the best available evidence in critical policy domains are legitimate.[110] Satisfying this standard may entail elevating the status or power of experts in the political process by, for instance, affording them a salient consultatory role or even some kind of veto power over legislation. Perhaps constitutional-based judicial review provides a helpful model here; one can imagine a "Supreme Court of Climate Experts" tasked with evaluating, modifying, or striking down legislation to the extent it exacerbates the climate crisis or contributes to other grave forms of environmental destruction.

The third, and perhaps most obvious, emerging CL factor is the *sustainability standard*. Sustainability is a difficult concept to define.[111] But the idea here would be that those governments which act so as to preserve and protect the natural environment, domestically and abroad, are (more) legitimate, while those that do not, are not (or are less so). This standard may conflict with a full-throated provision of individual rights, insofar as ensuring sustainability may sometimes mean limiting the extent to, or manner in, which individuals can interfere with other species or particular ecosystems. Relatedly, if we accept, as many compellingly argue,[112] that carbon-fueled, growth-centric capitalism is a root cause of climate change and inherently unsustainable, then the sustainability standard may also entail that those governments which refuse to transcend "fossil capitalism"[113] are, for that

[106] Extinction Rebellion 2022.
[107] See e.g. Holden 2002; Eckersley 2004; González-Ricoy and Gosseries 2016.
[108] Caney 2016.
[109] Grandoni 2022.
[110] See Frederiksen 2022 for an interesting discussion of how perceptions of competence can increase support for (even undemocratic) leaders. See also Di Paola and Jamieson 2018: 394ff for a discussion of how voter ignorance and a lack of governmental expertise cut against democratic legitimacy.
[111] Thiele 2013.
[112] The literature here is voluminous, but see e.g. Barry 2012, 2021; Klein 2014; Malm 2016.
[113] Malm 2016.

reason, illegitimate (or at least *less* legitimate).[114] In any case, ensuring some viable notion of sustainability appears increasingly relevant to normative assessments of political legitimacy, and perhaps, with time, will become even more salient than many of the individual freedoms and rights so cherished today.

This is only a cursory treatment of complex emerging trends in public morality. But there can be little doubt that these three standards are already shaping public perceptions of political legitimacy, and will likely only grow more salient as the crisis deepens. Evidence of this is the extent to which civil disobedience campaigns—which, as Chapter 4 argues, can plausibly be interpreted as responses to illegitimate political power—often describe their work in just these terms: i.e. as demands for greater stewardship, to heed the words of experts, or to implement sustainability policies. Yet the extent to which these emerging standards will come to supplant, or at least condition, prevailing CL standards is something that can only be discerned fully in retrospect.

3.6 Conclusion

In this chapter, I have argued that political legitimacy operates on two distinct levels. Most fundamentally, legitimacy requires that governments ensure the safety and security of their citizens, now and into the future; governments that cannot or will not perform this function are, for that reason, illegitimate. Yet, while ensuring safety and security is a *necessary* condition of legitimacy, it is not all that matters—at least not normally. Legitimacy also requires that governments exercise their power in broadly acceptable ways. What counts as acceptable, however, varies considerably across generational and cultural lines, and in light of different circumstances. For this reason, even the most widely celebrated standards—including the protection of basic rights and adherence to democratic processes—remain open to contestation and revision, and in this sense are contingent.

Under normal, reasonably favorable conditions, these two modes of legitimation operate in harmony. In crisis moments, however, prevailing liberal-democratic CL factors may obstruct a government's ability to preserve or restore conditions of safety and stability, and so ensure FL. In these cases, legitimacy may necessitate recourse to emergency powers, which are often authoritarian in character and scope. This is undoubtedly an uncomfortable conclusion. Yet it suggests a valuable lesson: often, our best means for avoiding legitimating authoritarian power is to prevent or expediently redress the kinds of grave security threats that make the use of such power necessary.

[114] On this, see e.g. Barry and Eckersley 2005; Di Paola and Jamieson 2018; Eckersley 2023; McGeown 2023, esp. ch. 2.

We may be too late for this with respect to climate change, however, which, after decades of inaction, represents a clear threat to public safety across the world (albeit much more so in some regions than in others, at least in the near term[115]). With little time left, only a rapid—and thus potentially rights- and democracy-abridging—mobilization of resources may be sufficient to prevent its most (politically) catastrophic effects. Of course, if governments can undertake action within the purview of prevailing CL factors—which is to say, *without* infringing upon individual rights or democratic processes—they should do so. But if adhering to CL factors proves incompatible with responding effectively to the climate crisis, then political legitimacy may require adopting a more authoritarian approach.

[115] Poor and developing states are often among the most vulnerable, both for geographical and economic reasons (Posner and Weisbach 2010: 21–3).

4
Disobedience in a Climate of Necessity

> New oil and gas is criminal, an act of genocide against billions of people in the poorest countries on earth, and an act of war against the young. We are not prepared to just watch while everything we love is destroyed. We are going to stop new oil and free ourselves from fossil fuel extortion and destitution, whether those in power agree or not.
>
> Just Stop Oil (2022)[1]

4.1 Introduction

On a cloudless May morning in 2014, Jay O'Hara, a Quaker from Cape Cod, and Ken Ward, Deputy Director of Greenpeace USA and former President of the National Environmental Law Center, prayed together before boarding a 32-foot lobster boat, dubbed the "*Henry David T.*," and navigating to the middle of a busy coal shipping channel near Brayton Point in Narragansett Bay. After dropping the anchor, Ward and O'Hara informed the local Somerset police of their intention to impede the *Energy Enterprise*, a ship bearing 40,000 tons of Appalachian coal. Once in sight, the *Energy Enterprise*'s crew hailed the *Henry David T.*, demanding they clear the way. When Ward and O'Hara refused, the *Enterprise*'s exasperated captain spit back, "But this is American coal!" About five hours later, the US Coast Guard boarded the *Henry David T.* and arrested the two activists. The *Energy Enterprise*, however, was unable to dock for the remainder of the day—a costly delay.[2] Criminal charges were filed against O'Hara and Ward; both were spared later that year, however, when, in a dramatic moment, the prosecuting attorney joined the activists on the front steps of the courthouse to announce he was dropping all charges against them.[3]

I argue that this incident—which we can call the Lobster Boat case—is representative of an important class of civil disobedience, which has so far not been very well examined, despite its potential political and legal significance.[4] In fact,

[1] Just Stop Oil 2022c.
[2] Lobster Boat Blockade 2013, 2014; Stephenson 2017.
[3] Bidgood 2014.
[4] At least by political theorists, although in recent years this has started to change; see e.g. Garcia-Gibson 2022; Scheuerman 2022; Lai and Lim (forthcoming). There are also some earlier exceptions: e.g. Welchman 2001; Lemmons and Brown 2011; Booth 2012.

Climate Change as Political Catastrophe. Ross Mittiga, Oxford University Press.
 DOI: 10.1093/oso/9780192868879.003.0004

while voluminous literatures exist on both disobedience and the political ethics of climate change, the intersection of these areas—climate disobedience—has largely been overlooked.[5]

Extending critical analysis here is important for at least two reasons. For one, as climate disobedience continues to grow in scale, frequency, and intensity across the world, it is coming to play an increasingly salient role in public conversations and governmental decision-making about climate change. Thus, understanding contemporary climate politics requires taking climate disobedience seriously.

Moreover, I argue that actions like the Lobster Boat case present a significant challenge to prevailing theories of civil disobedience. The most significant issue concerns the ends of disobedience. Notably, for all the debate and disagreement in the literature, there exists a virtually universal consensus that civil disobedience is fundamentally a *communicative* act.[6] According to this view, the basic goal in undertaking a disobedient act is to express one's moral convictions, expose and condemn injustice, and/or persuade the broader community to join the opposition. Without denying that many cases of climate disobedience fit this mold, an essential argument of this chapter is that others—those like the Lobster Boat case—are not (primarily) communicative. Rather, as the climate disobedients involved have themselves repeatedly insisted, the real end of their action is the *prevention of harm*—above all, the very great harm of politically catastrophic climate change.[7] If this is right, then the consensus view has missed something critical: civil disobedience need not be construed as (purely) communicative; it can (also) be harm-preventative.

In section 4.2, I further elaborate and defend the distinction between communicative and harm-preventative forms of civil disobedience. (Hereafter, I refer to communicative civil-disobedience as C-CD, and harm-preventative civil-disobedience as HP-CD.) One subtle but critical difference, I claim, concerns the way in which each instrumentalizes the penal process. A common view in the literature is that, while voluntarily accepting punishment is not an essential feature of civil disobedience, it may nonetheless prove a strategically valuable means for furthering the communicative ends of the protest.[8] This logic breaks down, however, if the aim of disobedience is not communication but the prevention of harm. Given this, we might expect that climate disobedients would typically attempt to evade punishment. And in some sense this is often the case. For when faced with criminal charges many have pleaded not guilty on the grounds that they acted

[5] There is, however, some important work exploring this subject in legal studies, as discussed in section 4.2.4.

[6] On this see e.g. Delmas 2018: 36–7. I return to this point in section 4.2.2.

[7] In this sense, as I explain in section 4.2.3, civil disobedience can have the aim of fulfilling the core precautionary duties discussed in Chapter 2, particularly when governments fail to do this themselves.

[8] Notably, even those who appear to think that accepting punishment *is* essential often acknowledge this instrumental value. I discuss this further in section 4.2.2.

out of "necessity."[9] Paradoxically, however, many of those same disobedients have reported themselves to the authorities and patiently awaited arrest, and some have described prosecutors dropping charges against them as "disappointing."[10] If they simply wished to avoid punishment, what explains this?

The key to resolving the paradox, I argue, is to understand that while these disobedients are not instrumentalizing punishment, they are instrumentalizing the penal process. For in pleading "necessity," they work to establish a legal precedent for regarding disobedience as an excusable response to climate harms and the broader threat of catastrophe. Such a precedent would have the effect of empowering others to take similar action against harmful actors by affording them greater assurance that their disobedience will not be met with jail time. And through this proliferation of action, climate disobedients endeavor to reclaim a form of public authority over agents and practices that threaten serious harm. We can see this strategic logic in play by examining particular cases of climate-related HP-CD and the professed motivations of those who undertook them, which comprises the focus of section 4.3.

Of course, successfully arguing necessity requires establishing that one's illegal act was, in fact, *necessary* for stopping or preventing harm. This suggests a powerful objection, taken up in section 4.4: that it is highly implausible, if not altogether incoherent, to claim that any single instance of (especially *civil*) disobedience could possibly be "necessary" for preventing climate harm, at least in any perceptible sense. However noble such actions may be, they are ultimately no more significant than single drops of water in an ocean. If this is right, then regarding climate disobedience as a genuinely *harm-preventative* act would be misguided.

In responding to this objection, I turn to the surprisingly analogous debate over the significance of individual votes in mass elections. In broad terms, I argue that *so long as* a particular act of climate disobedience serves to bring us closer to meeting critical emissions-reduction targets, and *so long as* the world ultimately succeeds in reaching those targets and therefore averting catastrophe, *then* there is a coherent basis for regarding the particular disobedient act, and any like it, as having been causally necessary to achieving that end. Thus the use of the necessity defense in such cases is legitimate.

Section 4.5 takes up additional objections concerning: (i) whether the substitutability of disobedient actions undermines claims of necessity; (ii) the likelihood that we will reach collective emissions-reduction targets, such that particular acts of HP-CD will, in fact, have been necessary; and (iii) opportunity costs and the comparative efficacy of HP-CD vis-à-vis other potential forms of individual climate action. Section 4.6 then concludes the chapter, offering some reflections over

[9] In broad terms, the necessity defense is a formal juridical claim that aims to excuse illegal actions undertaken to prevent exigent and more severe harms. I elaborate this below.

[10] Associated Press 2018; Hand 2018.

the relationship between HP-CD and more extreme forms of political resistance, which I claim are likely to become more prevalent if the existential threat of politically catastrophic climate change continues to be met with public disinterest and governmental inaction.

Before getting started, two clarifications are necessary. First, in discussing "prevailing" theories of civil disobedience, I do not mean to imply more scholarly agreement than exists. There are many contentious points in the literature, as I discuss later in this chapter. That said, it is not necessary for my argument to show that climate disobedience presents a *comprehensive* challenge to *all* theories of civil disobedience; rather, I only defend the more modest claim that a broad range of existing theories have overlooked HP-CD, and thus are unable to elucidate the distinctive features and potential significance of cases of climate disobedience which fit that mold. Second, in speaking of "climate disobedience" throughout, I should be understood to be referring specifically to cases that are undertaken in civil ways, and which aim, directly and in the first instance, at preventing climate harm. This of course is not to deny the existence of uncivil and/or communicative forms of climate disobedience, of which there are many examples.[11] I restrict my focus here to HP-CD cases, however, because I believe these to have a special legal and political significance, as I intend to demonstrate below.

4.2 Distinguishing between communicative and harm-preventative civil disobedience

4.2.1 Civil disobedience in general

There exists considerable debate over the defining features of civil disobedience. Among the earliest and perhaps still most influential accounts are those developed by liberal thinkers in the 1960s and 1970s, as an attempt to understand the US civil rights movement and its relationship to the rule of law and democratic order.[12] For these thinkers, civil disobedience is typically understood to be an illegal action, undertaken conscientiously, publicly (i.e. without an attempt by the disobedients to obscure their identities or evade authorities), nonviolently (i.e. without causing bodily harm or property damage), and with a willingness to accept legal punishment.[13]

Over the last few decades, however, scholars have identified significant disjunctures between this "orthodox" liberal conception of civil disobedience and important real-world cases. This has led many to call for relaxing, redefining, or

[11] I attend to these other forms in a paper entitled "Two ways of parsing disobedience" (Mittiga unpublished).
[12] Pineda 2015; Scheuerman 2015: 6n.21; Delmas 2018: 7, 21–4; Livingston 2018.
[13] See e.g. Bedau 1961: 661; Rawls 1999: 320.

simply abandoning some of the liberal conception's signal features in favor of a more capacious and so inclusive account. For example, some philosophers, like Marshall Cohen or Robin Celikates, have challenged restrictive interpretations of the nonviolence criterion, arguing, among other things, that property damage should not count as violence.[14] Perhaps more ambitiously, Kimberly Brownlee has contended that to "satisfy the non-evasion criterion" civil-disobedients "need not act to *ensure* that they are punished"; "it is enough," she maintains, that they "act communicatively and so take the risk of being arrested and punished or socially censured."[15] Joseph Raz, David Lefkowitz, and others similarly reject the idea that civil disobedience *must* involve the acceptance of punishment.[16]

There are many debates beyond these, which we can set aside here. It suffices to assume, with Delmas and Brownlee, that a given illegal act will be more or less *civil* to the extent that it comports with the standard norms of civility (publicity-in-action, nonviolence, etc.),[17] accepting that there will always be some interpretative latitude in how we understand those norms and their practical relevance in any particular case.[18]

4.2.2 The consensus view

Debate and disagreement notwithstanding, there exists a surprising consensus in the literature that civil disobedience is, at bottom, a *communicative* act. We can find this basic assumption expressed in many ways. Rawls, for instance, stresses that in undertaking civil disobedience "one addresses the sense of justice of the majority ... and declares that in one's considered opinion the principles of social cooperation among free and equal men are not being respected."[19] *Addressing* the community and *declaring* one's opinion through the illegal act are the ways in which disobedients build the majority support necessary for realizing political or legal change, and both are clearly communicative endeavors.[20] Brownlee's revisionist account—which, on many other points, is deeply critical of Rawls's—takes a strikingly similar view, defining civil disobedience as "a deliberate breach of law

[14] Cohen 1969, cited in Scheuerman 2019: 16n.9. See also Celikates 2013, 2016, esp. 985; Adams 2017: 488; Delmas 2018: 36.

[15] Brownlee 2012: 38.

[16] Lefkowitz 2007: 226; Raz 2009: 265; Celikates 2016: 985. Interestingly, Celikates's argument is developed in relation to the case of Edward Snowden, which has become a flashpoint in the literature on disobedience, especially with respect to the question of punishment. See e.g. Scheuerman 2014; Pineda 2015; Delmas 2018: esp. 55, 100–4.

[17] Delmas 2018: 42–4; Delmas and Brownlee 2021.

[18] Consider e.g. Celikates's (2016: 985) defense of evading punishment when one cannot expect a fair trial.

[19] Rawls 1999: 320.

[20] Rawls (1999: 321) later claims that, like a "public speech," civil disobedience is "a form of address, an expression of profound and conscientious political conviction, [that] takes place in the public forum."

taken on the basis of steadfast personal commitment in order to *communicate* our condemnation of a law or policy to a relevantly placed *audience*."[21] Brownlee specifies further that civil disobedience is communicative in two ways: (1) it expresses "our disavowal of, and dissociation from, the protested law or policy as well as the reasons for our disavowal," and (2) it "draw[s] attention to the reasons for the protest so as to persuade the relevant audience to accept our position and, thereby, to instigate a lasting change in law or policy."[22] Both of these two dimensions can be found in canonical accounts of civil disobedience. For instance, in "Letter from Birmingham Jail," Martin Luther King, Jr. claims that civil-disobedients act to "expose" injustice, "arouse the conscience of the community," and force open "the door to negotiation."[23] The key point here, as Delmas observes, is that a very wide range of theories—from the most traditional to the most radical—"maintain Rawls's core insight that civil disobedience is essentially a communicative act."[24]

4.2.3 The end of harm prevention

While this consensus view certainly holds with respect to a wide range of cases, I argue nevertheless that we should reject the rigid identification of civil methods with communicative ends. Civil disobedience may also have the end of impeding agents, institutions, or practices that are causing or contributing to serious public harms (or the credible risk thereof)—i.e. the end of fulfilling the core precautionary duties described in Chapter 2—particularly when governments fail to do so.

This harm-preventative form of civil disobedience shares many of the core features and commitments of its communicative counterpart: i.e. HP-CD activists operate *publicly*, without any attempt to hide their identities or evade authorities; their actions are *nonviolent*, at least in a reasonably capacious sense of that term;[25] they are *conscientiously* motivated, seeking, as they do, to combat grave threats to public safety in the absence of governmental action (or in the face of governmental malfeasance); and while the *acceptance of punishment* is not integral to HP-CD, for reasons already intimated in the introduction and described further later, acceptance of arrest and a similarly instrumental use of the penal process is.

[21] Emphasis added; Brownlee 2012: ch. 1. See also Brownlee 2017, which argues that civil disobedience centrally involves "a conscientious or principled outlook and the communication of both condemnation and a desire for change in law or policy."

[22] Brownlee 2012: 18–19.

[23] King 1992.

[24] Delmas 2018: 36–7, 18.

[25] In other words, HP-CD activists strictly avoid physically harming other people or nonhuman animals, although they may often cause property damage or otherwise impose financial costs on harmful actors, to the extent necessary for preventing harm.

HP-CD is political in various senses. For one, it is not oriented toward reprisal for harms done or aiding victims of said harms,[26] but rather toward inhibiting ongoing or future harms, until the relevant structures change or culpable agents desist. This is especially important in the context of two basic kinds of political failure:

(1) *Endangerment by action*: When a government inexcusably[27] causes or facilitates serious public harms, or the credible risk thereof.
(2) *Endangerment by omission*: When a government inexcusably fails to prevent or remedy serious public harms, or the credible risk thereof.

Endangerment in either (or both) of these respects undermines "foundational legitimacy" (FL). Recall from Chapter 3 that, in order to possess FL, a government must ensure the basic safety and security of its citizens, now and into the future. To the extent a government fails at this essential task, it is rightly considered illegitimate. And political illegitimacy, should it become sufficiently grave and enduring, is enough to justify principled departures from the rule of law—i.e. disobedience. Indeed, such illegitimacy may even *necessitate* disobedience (in the mode of HP-CD[28]); for, if a government is unwilling or unable to act,[29] then it falls to the people themselves to serve as the final guarantors of safety, even if this means operating outside of, or in violation of, the law.

Interestingly, some theorists appear to have recognized that at least some instances of political disobedience are not communicative, though this claim has not been well developed. Moreover, among those who have identified this possibility, there is a pronounced (and, I claim, misguided) tendency to associate the pursuit of noncommunicative ends with the use of uncivil tactics. In other words, just as the consensus view (wrongly) presumes that all instances of civil disobedience are communicative, these theorists often (wrongly) assert that all instances of noncommunicative disobedience are uncivil.

Delmas provides a clear example. She perceptively observes that "[s]ome acts of uncivil disobedience are primarily communicative, but many do not aim to persuade an audience—seeking instead to prevent or redress wrongs—though they may still include communicative elements."[30] A few pages later she elaborates this idea, claiming that "[n]oncommunicative uncivil disobedience, such as direct

[26] In this sense, disobedience is distinct from vengeance or retaliation, as well as from relief work (Delmas 2018: 93–4), or what Brownlee (2012: 24–5) calls "assistive disobedience."

[27] See Chapter 2, section 2.5.2.1.

[28] Or rather harm-preventative disobedience in general, which includes uncivil forms. On this, see Mittiga (unpublished).

[29] Which, again, is to say: if a government is unable or unwilling to fulfill its core precautionary duties, described in Chapter 2.

[30] Delmas 2018: 44–5. See also Delmas 2018: 52–3, 59, 63, 94. Compare to similar remarks in Delmas and Brownlee 2021: "Uncivil disobedience ... can be used to designate acts of principled disobedience that *may or may not be communicative*" (emphasis added).

action" is typically "justified" when "the state endangers or harms some people or unjustly fails to protect them from harm."[31]

These observations merit further development. As a first point, note how the terminology Delmas employs—"noncommunicative" disobedience[32]—is suggestive of the strength of the consensus view, in that ends beyond communication are referred to simply in the negative, as exceptions to a general rule. More relevant for our purposes, though, is the strict association Delmas implies between the end of harm-prevention and the use of uncivil means, as well as her conflation of harm-preventative disobedience with the concept of "direct action." These two issues are closely related, as direct action is routinely theorized as a *definitionally* uncivil form of protest aimed at thwarting objectionable activities or actors, rather than communicating dissent. William Smith evinces this tendency in defining direct action as

> a mode of disruptive activism carried out with the aim of obstructing or deterring contentious practices that are opposed by activists. This form of action is "direct" because it is an attempt to *stop a perceived wrong from occurring*, rather than simply an attempt to *communicate* opposition to that wrong ... The tactics that have been associated with direct action include open or covert attempts at disruption, such as blockades, trespass, occupations, sabotage, property damage and other types of force.[33]

For Smith, direct action's noncommunicative (i.e. harm-preventative) ends, along with its "less constrained" or even "militant forms of conduct," place it outside the realm of civil disobedience, properly understood.[34] According to the argument I am advancing here, Smith's view is confused insofar as it fails to distinguish between ends and means. Doing this, we are able to see that preventing harm does not imply any particular mode of intervention, be it civil or uncivil.

Smith is of course not alone in this. We can find the same basic tendency throughout the literatures on civil and uncivil disobedience and direct action. Scheuerman, for example, explicitly defines a strain of direct action—what he calls "blocking and disrupting disobedience" (BD)—against the concept of "nonviolent civil disobedience" (NVCD), writing:

> BD obviously breaks with core features of NVCD. Since the former does not conceive of lawbreaking as a form of *communicative* address aimed at the public

[31] Delmas 2018: 52, 53.

[32] Notably, the same terminology ("noncommunicative") is used by Brownlee (2012: 27), though for her, noncommunicative disobedience is essentially individual non-compliance, closer in spirit to what most refer to as conscientious objection.

[33] Emphasis added; Smith 2018: 13.

[34] Smith 2018: 25, 14.

> and/or state officials, it can be *uncivil* (e.g., secret, legally evasive, and even violent).[35]

Celikates likewise conceptualizes direct action as a form of disobedience that employs what would standardly be considered uncivil methods, in pursuit of ends that "go beyond ... moral appeal."[36]

It is important, then, to clarify the relationship between direct action and what I am calling harm-preventative disobedience. Insofar as direct action aims, primarily and in the first instance,[37] at preventing significant public harms, it clearly coincides with harm-preventative disobedience. Yet, if we understand direct action, as many appear to do,[38] as a *necessarily* uncivil form of disobedience, then we must not take it to be exhaustive of harm-preventative disobedience, which, again, can be either civil or uncivil. In other words, these assertions miss the basic point that the end of harm-prevention is perfectly compatible with civil tactics. This oversight is problematic for more than purely conceptual reasons. As I will argue, HP-CD has a particular political and legal significance, which we can only appreciate if we are able to distinguish it properly from other forms: i.e. civil and uncivil communicative disobedience, as well as uncivil harm-preventative disobedience.[39]

A clear example of HP-CD is discussed in one of the earliest general treatments of civil disobedience—that of Hugo Bedau, who describes the voyage of the ship *The Golden Rule*, which:

> in the spring of 1958 [sailed] into the Central Pacific, to try to force the United States government to abandon its nuclear-weapons testing program or to deliberately expose several of its citizens to a probably fatal dosage of radioactive fallout, since the crew intended to sail the ship directly into the testing area.[40]

The crew of *The Golden Rule* never reached their destination, having been apprehended en route by the US Coast Guard, charged with contempt of court, and subsequently sentenced to sixty days in jail. The critical point, though, is that

[35] Emphasis added; Scheuerman 2022: 802.

[36] Celikates 2016: 984. Notably, though, even in recognizing these features—the tendency to use what are widely regarded as uncivil tactics (such as sabotage, blockades, and direct confrontation) and the pursuit of ends apart from communication—Celikates nevertheless attempts to assimilate direct action into his account of civil disobedience, which works only to the extent that one accepts his implausibly capacious notion of civility. Of course, Celikates is not alone in this; see e.g. Welchman 2001.

[37] Which is not to say *exclusively*.

[38] In addition to the works already cited, see also Bedau 1961: 357. Of course, there are exceptions. Celikates (2016), for instance, seems to think that many forms of direct action are civil, but as noted above he adopts an unusually expansive (and, for that reason, as Delmas [2018: 36] claims, almost incoherent) conception of "civil."

[39] Again, for further discussion of these different forms see Mittiga (unpublished).

[40] Bedau 1961: 657.

these activists broke the law with the primary aim of preventing real harm *while conforming* to the standards of civility—a model of disobedience adhered to by many subsequent antinuclear activists.[41] In section 4.3, I discuss other examples of HP-CD, particularly from the climate movement.

4.2.4 Instrumentalizing the penal process

First, though, we must note one point of difference between C-CD and HP-CD: the way in which each seeks to exploit the penal process. This difference is subtle but nonetheless critical, particularly with respect to recent climate cases.

Of course, as noted earlier, there is widespread disagreement in the existing literature over whether civil-disobedients ought to willingly accept punishment. Nevertheless, many—even among those who do not believe that activists have a moral duty to accept punishment—regard this as a valuable way of furthering the communicative ends of the protest. Rawls, for instance, argues that disobedients need not regard their punishment as morally "right," but they should accept that, when they challenge particular laws or practices, they "must pay a certain price to convince others that [their] actions have ... a sufficient moral basis in the political convictions of the community."[42] David Lefkowitz makes a similar point, albeit in more straightforwardly political terms. He argues that the acceptance of "temporary confinement by the state" allows disobedients to communicate to their fellow citizens that they "are not trying to usurp [democratic] authority"[43] by "coercing the state into abandoning or adopting certain policies,"[44] but "instead are acting (just) within the boundaries of political debate."[45] King also focuses on the instrumental value of "willingly accept[ing] the penalty" of civil disobedience, which, he argues, can help "arouse the conscience of the community," while "expressing the very highest respect for law."[46] In other words, for King, the value of accepting punishment comes from exploiting the fructive moral tension that arises when a just person is punished for violating an unjust law. Forcing a community to confront this tension can generate sympathy for disobedients (and by extension their protest), thereby helping to persuade others to join the opposition. Ten-Herng Lai has this basic idea in mind when he argues that

> punishment plays an indispensable role in contributing to civil disobedience's overall ability to serve as a *costly signal*: one that allows civil disobedients

[41] See e.g. Aldridge and Stark 1986; Lawrence 1988. Further examples are considered later in this chapter.
[42] Rawls 1999: 322. Bedau (1961: 659) also makes something like this point. But cf. Livingston 2018.
[43] Lefkowitz 2007: 222.
[44] Lefkowitz 2007: 216.
[45] Lefkowitz 2007: 222.
[46] King 1992.

> to distinguish themselves from speakers who lack the relevant sincerity and seriousness, which in turn constitutes the reliability and effectiveness of their communicative act.[47]

For this reason, Lai contends, "[w]ithout punishment, the effectiveness of civil disobedience is compromised." The key point here is that, for many thinkers, accepting punishment is considered instrumentally valuable in that it contributes to the success of civil disobedience as a communicative act.[48]

As with C-CD, many of those involved in HP-CD regard publicity-in-action and the willing acceptance of *arrest* as core components of civility. Nevertheless, insofar as their primary aim is not communication but the prevention of harm, these same disobedients do not take the willing acceptance of *punishment* as a strategically valuable part of their protest. Quite the opposite: often, during criminal proceedings HP-CD activists make use of a particular legal mechanism, the "necessity defense," which, when employed successfully, excuses their action and so spares them criminal punishment. Evading punishment is not the (primary) aim, however. As John Alan Cohan explains, in using the necessity defense in such cases: "[a]cquittal is of course hoped for in the end but may be quite low on the protestors' list of priorities."[49] For HP-CD activists, the real point in pleading necessity is, I argue, to establish a legal precedent that would treat their and like actions as excusable responses to dire threats to public safety. In this way, they instrumentalize the penal process, but not punishment.

Before elaborating this point, it is helpful to clarify the basic form of the necessity defense and what must be established in order to employ it. Typically, defendants must demonstrate that the following conditions were met at the time of their illegal action:[50]

(1) There was a reasonable expectation of imminent harm.
(2) The imminent harm exceeded that which was caused by breaking the law, and in this sense the illegal act constituted the "lesser of two evils."
(3) The imminent harm was not created by the person or people breaking the law to prevent it.

[47] Lai 2020: 1084.
[48] Note that some who accept the instrumental value of accepting punishment do so on different grounds. Livingston (2018: 527), for example, argues that, for Gandhi, accepting punishment follows from a "fidelity to truth" rather than to law, and functions foremost as a means of self-purification through enhancing the cultivation of the "virtues of non-violence."
[49] Cohan 2007: 111.
[50] With the exception of (3), these are the conditions set out in the landmark case *United States v. Maxwell* (Long and Hamilton 2018: 59n.14).

(4) The defendants reasonably believed that there existed a "direct causal relationship" between their actions and the prevention of the harm.
(5) No reasonable legal alternatives to breaking the law were available.[51]

A paradigmatic (nonpolitical) example of a necessity case is as follows:

> Suppose that your neighbor's house is on fire and that if you do not act within a few minutes the entire block will be consumed. Your legal option is to call the authorities and to wait for them to arrive, at which point it will be too late. Your illegal option is to break into your neighbor's home and extinguish the fire yourself.[52]

Under these conditions, disobedience is considered excusable[53] on consequentialist grounds: should one adhere strictly to the letter of the law and fail to intervene, substantially greater harm will occur than would have if the illegal act had taken place.[54]

A range of HP-CD activists over the last fifty years have attempted to make use of this defense in explicitly political cases, with varying degrees of success. This trend was especially pronounced in the 1960s and 1970s, so much so that legal scholars began referring to "the political necessity defense" (PND).[55] Appeals to the PND were most common among antinuclear and antiwar protestors in the US.[56] Consider, for instance, *State v. Mouer*, an Oregon case in which eighty-two activists of the "Trojan Decommissioning Alliance" stormed a local nuclear power plant in an attempt to terminate its operations. This lasted two days, at which point all were arrested and charged with second-degree criminal trespassing.[57] Another example is *United States v. Berrigan*, in which two priests (Philip and Daniel Berrigan) broke into a federal Selective Service office and poured blood over the draft records to stymie conscription there. In the first case, the PND was successful; in the second it was not.

It exceeds my purposes to provide a more extensive discussion of the history of, or debates surrounding, the use of the PND. The important point here is that, as

[51] While these five conditions are common, there is some variance in how they are formulated or ordered in different legal jurisdictions. Also, in some jurisdictions, certain conditions are omitted (e.g. (3)) or additional ones added (e.g. that the defendant ceased committing the illegal activity as soon as the harm was averted). For further discussion, see Long and Hamilton 2018 and Rausch 2019. See also Climate Defense Project 2021.

[52] Quoted in Long and Hamilton 2018: 99.

[53] See Ripstein (1998: 163–70) for a helpful discussion of the difference between "excuses" and "justifications" in a legal context.

[54] For a defense of the claim that emergency conditions can necessitate a turn to consequentialist ethics, see Adams and Mittiga 2021.

[55] The relevant literature here is vast, but see e.g. Aldridge and Stark 1986; Bauer and Eckerstrom 1986; Lawrence 1988; Cohan 2007; Rausch 2019. For a political-theoretic analysis, see Brownlee 2012: ch. 6.

[56] As noted in Long and Hamilton 2018 and Rausch 2019.

[57] Aldridge and Stark 1986: 310.

these cases demonstrate, by accepting arrest but subsequently seeking to excuse their actions by appeal to necessity, HP-CD activists are exploiting the penal process in a distinctive but nonetheless instrumental way. Again, unlike C-CD, the aim is not to use criminal hearings as a means of signaling credibility or generating sympathy, but rather to establish legal precedents that could embolden others to take similar action against harmful agents and practices.

Of course, achieving this requires HP-CD activists to expose themselves to serious legal risk by leaving their fates to be decided by juries. Yet, in this way, HP-CD activists affirm democratic authority[58] and express something like an inverted fidelity to law—not to the strict letter of legal codes, but rather to the ineliminable element of living human judgment contained in the interpretation and application of law.[59] In effect, when juries validate the necessity of disobedience, they open up space for similar future acts of HP-CD, and thus for revitalized public agency over the forces that threaten the well-being of the polity.

4.3 Harm-preventative civil disobedience in the climate movement

We can better understand HP-CD—and its potential legal and political significance—by looking at some recent high-profile cases in the climate movement, beyond the Lobster Boat case described in the introduction. Again, to be clear, I restrict my focus here to examples in which those involved (i) explicitly described their actions as civil, and adhered to (most of) the standard norms of civil disobedience (e.g. nonviolence, publicity-in-action), but (ii) nevertheless aimed, directly and in the first instance, at the prevention of harm.

The first example is an action that took place on December 19, 2008, when Tim DeChristopher traveled to a mineral-rights auction conducted by the US Bureau of Land Management with the intention of protesting outside. After arriving, however, DeChristopher decided to enter, and, once inside, falsely identify himself as a bidder. He then proceeded to substantially inflate the prices of many oil and gas leases and outright won fourteen, worth $1.8 million[60] and spanning nearly 22,000 acres.[61] Soon, however, others in the room became suspicious and reported DeChristopher, leading authorities there to detain him. A few months later, the Obama administration deemed the auction illegal and cancelled all awarded leases. Charges were nevertheless brought against DeChristopher, resulting in him being forced to serve nearly two years in jail.

[58] After all, there is perhaps no more democratic an institution in the US than juries.

[59] While I cannot pursue the point here, this suggests a parallel with emergency authority, which also exceeds the letter of the law for reasons of necessity, thereby affirming the value of that which is to be saved. On this point, see Lazar 2009.

[60] Maffly 2018.

[61] Bidder 70 2012.

Another case, which garnered considerable media attention and has since influenced many similar acts, is that of the "Valve Turners." After months of safety preparations, on October 11, 2016, Emily Johnston, Annette Klapstein, Leonard Higgins, Michael Foster, and Ken Ward (of the Lobster Boat case) simultaneously closed the valves of five major pipelines in four different states,[62] all of which were transporting (tar-sands) crude oil from Canada to the US. Associates filmed and live-broadcast their actions over social media. Within fifteen minutes of the valve closings, the activists called the relevant pipeline companies to announce what they had done, and subsequently awaited arrest. Their actions succeeded in blocking nearly 15 percent of the US's total crude oil imports for that day,[63] which Reuters called "the biggest coordinated move on US energy infrastructure ever undertaken by environmental protesters."[64] All of the activists (and several of their associates) faced felony-level criminal charges; however, as of August 2021, only one had been convicted while the others have been freed or are awaiting appeal.[65]

A third, even more recent, example is that of the Just Stop Oil campaign in the UK. Of the dozens of disobedient actions those involved with this campaign have undertaken, many have focused on impeding the distribution of fuel by blockading oil terminals, gas stations, and other "critical oil infrastructure." From April through August 2022, this resulted in serious fuel shortages throughout the Midlands and south of England.[66] Moreover, in at least one instance, delivery vehicles operating out of the Kingsbury oil terminal were targeted, "causing damage estimated at £184,000."[67] As of September 17, 2022, more than 1,350 Just Stop Oil disobedients have been arrested, and at least fifty-four have been remanded to prison.[68]

In all of these cases, and in many more like them, the activists involved have been at pains to comport with what they understood to be the core features of civil disobedience. For instance, they acted publicly, with no attempt to hide their identities or evade authorities (in many cases they reported themselves, or publicly declared their intentions in advance). They actively sought to avoid causing unnecessary property damage and, crucially, any physical harm to others (hence the Valve Turners' safety preparations and call to the pipeline companies, or the *Henry David T.*'s clear signaling of their intentions to the other ship and the relevant authorities). And yet, the end of these protests was not to express moral

[62] Specifically, Michael Foster stopped a TransCanada pipeline in Grand Forks, ND; Ken Ward "closed a shut-off valve on Kinder Morgan's Trans Mountain pipeline near Mount Vernon," WA; Leonard Higgins "closed a shut-off valve on the Enbridge Express Pipeline near Coal Banks Landing," MT; while Emily Johnston and Annette Klapstein "turned the shut-off valves on a pair of pipelines owned by Enbridge" in Leonard, MN (Nijhuis 2018).

[63] Civil Liberties Defense Center 2023.

[64] Williams and Kearney 2016.

[65] Climate Defense Project 2021.

[66] Gayle 2022a, 2022b.

[67] BBC News 2022.

[68] Taylor 2022.

convictions, expose injustice, or elicit sympathy but, unequivocally, to stop—as far as those involved were able—significant sources of climate harm. They accomplished this in various ways: directly, by, for instance, preventing the use of fossil fuels, disrupting supply lines, damaging existing and slowing the construction of new infrastructure; but also, indirectly, by forcing costly delays, imposing financial losses and damages, and creating new forms of risk for harmful agents, which make extracting, processing, and distributing fossil fuels logistically more difficult and financially less lucrative.[69]

That these activists have aimed primarily at preventing harm is clear from their public remarks. Reflecting on the Lobster Boat case, for instance, Ward and O'Hara explained that their intention was straightforwardly to block fossil fuels from being burned in New England, which was necessary given the absence of meaningful political action.[70] They later elaborated their reasoning:

> [E]very day of additional emissions is a terrible, immoral imposition on our children and, in ways we do not fully understand, on the other living things of God's creation ... It is our choice to take direct, non-violent action—putting our bodies between the Brayton Point coal plant and its water-borne coal supply—in an attempt to achieve the outcome necessary for planetary survival.[71]

Just Stop Oil has explained the motivation for its disobedience campaigns in much the same terms:

> Allowing the extraction of new oil and gas resources in the UK is an obscene and genocidal policy that will kill our children and condemn humanity to oblivion. It just has to stop. If we continue down our current path it will destroy families and communities. We will face the starvation and the slaughter of billions of the poor—and the utter betrayal of our children and their future.[72]

While on trial for participating in an oil terminal blockade despite a court injunction against further protest activity, one Just Stop Oil disobedient in particular—Zoe Cohen—said plainly but defiantly: "The judiciary is colluding with people destroying life on earth. I will break the injunction again—I have a moral duty to do so."[73]

In part because of the media attention they have received, and in part because of increasingly dense organizing and information-sharing within climate activist communities, these cases have proved highly influential and are now

[69] I thank one of OUP's anonymous reviewers for the point about HP-CD's indirect effects.
[70] That is, in response to the second type of government failure noted earlier. See Allen 2013; Lobster Boat Blockade 2014 (especially from 52 seconds onwards).
[71] Queally 2013.
[72] Just Stop Oil 2022b.
[73] Just Stop Oil 2022a.

representative of a larger array of disobedient actions, especially in North America, Australia, and Europe.[74] Uniting them is a belief that disobedience is a vital part of the broad collective effort needed to stave off climate catastrophe and the massive harms it entails. This is an essentially moral end,[75] but one undergirded by the existential fear that, at some point, the earth will no longer be capable of supporting human civilization—i.e. a fear of the politically catastrophic potential of climate change. Moreover, as with earlier instances of HP-CD, in many of these cases, activists have used the necessity defense (or its equivalent[76]) during criminal proceedings. In fact, the necessity defense is widely regarded as an *extension* of the protest—a further means to prevent harm. As the Climate Disobedience Center explains:

> Although the necessity defense employs the legal system of courts and judges, it is fundamentally a *political* and an *organizing* strategy that continues the direct action you took in the first place into a new phase ... It is a strategy for building movement power.[77]

And indeed, as one legal scholar argues, "if courts begin to accept the necessity defense" more widely, it will constitute "one tool, of many, that could be used to stop a complex and multifaceted issue such as climate change."[78]

No doubt for this reason, use of the defense has exploded in recent years: between 2008 and August 2021, there were more than thirty cases in the US alone in which climate disobedients pleaded (or attempted to plead) necessity, and this number grows every year.[79] This has been fueled by activist organizations and networks—like the Climate Disobedience Center, Extinction Rebellion, and the Civil Liberties Defense Center—which insist on the urgent need for disobedience to combat climate change, and which offer resources concerning how to argue necessity if prosecuted.[80]

When employing the necessity defense, disobedients have argued that decades of dilatory climate politics have allowed harms to multiply and the risk of catastrophe and collapse to grow. Moreover, the legal means available for enjoining lawmakers and other responsible agents to act are lacking, and time is running out. As O'Hara and Ward (of the Lobster Boat case) contend: "We are faced with

[74] For a helpful case guide, see Climate Defense Project 2020. For discussion, see Rausch 2019.
[75] Delmas 2018: ch. 5.
[76] The necessity defense goes by many names. In Montana, for instance, it is called the compulsion defense; in Oregon the "choice of evils" defense (https://cldc.org/climate-necessity-defense/); in the UK, it is often referred to simply as "lawful excuse" (Long and Hamilton 2018: 61–2).
[77] Climate Disobedience Center 2017a.
[78] Rausch 2019: 602.
[79] Climate Defense Project 2020.
[80] See e.g. the Climate Disobedience Center (https://www.climatedisobedience.org/necessity defense), Extinction Rebellion (https://www.xrebellion.nyc/legal-resources-protest), and the Civil Liberties Defense Center (https://cldc.org/climate-necessity-defense/).

an imperative like none confronted by any previous generation; we are living in a society that disavows responsibility for this greatest of crises and lacks any process or means by which decisions ... might be affected."[81] Just Stop Oil has put the same point in even starker terms:

> We are not prepared to just watch while they destroy everything we love. We're done with begging. Voting has changed nothing. We are going to stop new oil whether those in power agree or not. As citizens, as humans, as parents and children we have every right under British law to protect ourselves and those we love. This is the moment, we are the last generation that can solve this.[82]

In other words, these activists often regard disobedience as the *only* viable option available to them for stopping a serious and growing threat to collective safety and welfare: catastrophic climate change.

It may be tempting to regard appeals to necessity as just a roundabout way of evading punishment. And it is of course true that, in applying the necessity defense, climate disobedients seek to be found "not guilty." Yet, as with earlier HP-CD cases outside of the climate movement, avoiding punishment is at most a secondary consideration.[83] After all, if disobedients were using the necessity defense merely to evade punishment, why would they publicly broadcast their actions, report themselves to the authorities, or await arrest? Why not instead act in secret, fleeing the scene before police arrive? Moreover, how could we make sense of Valve Turner Emily Johnston's remarks after the court dropped all charges against her: "I ... admit that I am disappointed that we did not get to put on the trial that we hoped for"? (Notably, Johnston is not the only disobedient to have expressed disappointment at charges against them being dropped.[84])

The only plausible answer to these questions is that the activists involved were hoping to be exonerated specifically by means of the necessity defense, as this would—in the words of the Climate Disobedience Center's legal guide—set "a powerful precedent for future defendants to justify their moral lawbreaking."[85] For in setting such precedents, activists effectively reclaim a form of public authority over agents and practices that are causing serious harms and risking political catastrophe.

So far, this strategy has been met with only limited success, for various reasons. In the Lobster Boat case, for example, Ward and O'Hara were authorized

[81] Queally 2013.
[82] Just Stop Oil 2022a.
[83] As the Climate Disobedience Center's necessity defense guide stresses: "A necessity defense ... is not primarily a means of seeking an acquittal or a lighter sentence (though that *could* happen)" (Climate Disobedience Center 2017a).
[84] Associated Press 2018. Her fellow disobedient, Annette Klapstein, also described not having the chance to plead necessity as "disappointing," despite being set free (Hand 2018).
[85] Climate Disobedience Center 2017b.

to use the defense, yet the prosecutor dropped the charges against them before this could happen.[86] For the Valve Turners, who have faced separate trials in different states, appeals to necessity have been wholly or partially prohibited during initial proceedings,[87] and sometimes for nakedly political reasons. The judge in Johnston's case, for instance, exclaimed: "I'm not going to let you put US energy policy on trial!"[88] Likewise, the judge in Ward's first Valve Turner trial falsely claimed that "there's tremendous controversy over ... whether [climate change] even exists."[89] Such outcomes are fairly common: the prosecution dropping charges before necessity can be argued or the presiding judge barring the defense during a pretrial hearing. Nevertheless, as of August 2021, disobedients in twelve cases have had at least partial success in employing the defense.[90]

There are of course legitimate questions about the extent to which climate disobedience satisfies the formal requirements for arguing necessity. None of these potential doubts are irresolvable, however. On this point, much of what I have said in earlier chapters is relevant, particularly with respect to the imminent-harm and lack-of-legal-alternatives criteria.[91] Undoubtedly, though, the most challenging issue for climate disobedients is establishing that a "direct causal relationship" exists between their illegal actions and the prevention of harm: i.e. that (any one instance of) disobedience can plausibly be considered "necessary" in the fight against climate catastrophe. Examining this issue comprises my focus in the remainder of the chapter.

4.4 Does climate disobedience actually prevent harm?

To begin addressing this question, we should first note that, while there is certainly room for debate, legal scholars have argued compellingly for the appropriateness of the necessity defense in these climate disobedience cases.[92] And, as mentioned earlier, some of the most recent judicial decisions have been sympathetic to

[86] Long and Hamilton 2018: 62.

[87] In *Minnesota v. Klapstein*, however, the defense was partially allowed (in particular, some expert testimony regarding the dangers of climate change was admitted), before the judge simply dismissed all charges (Climate Defense Project 2020).

[88] Johnston 2017.

[89] Levin 2017; Long and Hamilton 2018: 66–7.

[90] Strikingly, most of these successes have occurred within the last five years, that is, since 2018. For more information on these cases, see Climate Defense Project 2021. For discussion of an early successful case, that of the "Kingsnorth 6," see Vidal 2008; Brownlee 2012: ch. 6; Long and Hamilton 2018: 61.

[91] Regarding imminent harm, see Chapter 3 section 3.4.1, and Chapter 2 section 2.6.1.2. On the lack of legal alternatives, see Chapter 3, which discusses the systematic failure of nearly all heavy-emitting states to combat climate change. There can be little doubt that the disastrously dilatory political machinery operating in most parts of the world has delivered us to this point. If effective legal alternatives exist, what are they?

[92] See e.g. Long and Hamilton 2018.

necessity claims, thanks in part to the testimony of expert witnesses—especially climate scientists like Jim Hansen, who have helped demonstrate the urgent need for action,[93] and civil disobedience scholars, who have provided historical evidence of the efficacy of direct action for creating political change. Partly for these reasons, in 2019 Ward succeeded in appealing the decision of his climate-change-denying judge on the grounds that barring the necessity defense violated his constitutional rights, which resulted in the Supreme Court of Washington establishing a right to use the necessity defense in all future climate disobedience trials.[94] Moreover, the difficulty of employing the necessity defense has not deterred activists from continuing to try, even as conservative legislatures have begun enacting increasingly draconian laws against common forms of harm-preventative climate disobedience, like valve closing.[95] Thus we have strong provisional grounds for thinking that the necessity defense is appropriate in these cases, and for expecting that it will be more successfully employed in the years ahead.

4.4.1 The objection

Nevertheless, it is not hard to see how one might object here that it is implausible to think that any individual act of climate disobedience—particularly *civil* climate disobedience—could play a causally significant role in preventing harm, let alone the very great harm of climate catastrophe. If this is right, then we would be wrong to regard climate disobedience as a genuine form of harm-preventative disobedience; however noble the intentions of those involved may be, the purpose of their illegal actions can only really be to draw attention to the crisis or, more crudely, give expression to their moral outrage. And while this may lead to the prevention of harm in a vague and roundabout way, that is not enough to regard climate disobedience as anything more than communicative or symbolic.

The strength of this objection is suggested by the fact that the few political theorists who have examined climate disobedience have studiously avoided attributing any material effects to it. For instance, for Scheuerman, part of the appeal of using the term "blocking-and-disrupting" disobedience (or BD) to refer to many of the same actions I have labeled HP-CD seems to be that BD is an explicitly procedural term (corresponding to the methods employed), which makes no allusion to actually preventing harm. This semantic hedging is indicative of the view that individual acts in the face of massive collective harms like climate change have no

[93] Mingle 2008; Long and Hamilton 2018: 61, 62.

[94] Corbett 2019; Climate Defense Project 2020.

[95] See e.g. Cagle 2019; Kusnetz 2019; Lefebvre and Adragna 2019. Some governments have also begun to designate environmental activist groups as "extremists," alongside neo-Nazis (Wilson and Walton 2019; Federman 2020); in the UK, however, this move was walked back after public outrage (Wall 2021).

perceptible impact. If this is right—if the objection holds—then the only possible point of participating in disobedience would be to telegraph the gravity of the situation and thereby enjoin public action; or, failing that, at least to shame and annoy the bad guys.

Note, though, that however much this objection comports with common-sense views of causality, it carries with it the uncomfortable implication that the many climate disobedients who have claimed to be acting directly to prevent harm, and who subsequently defended their actions by appeal to necessity, were either confused or engaging in legally opportune subterfuge. Perhaps some would wish to resist this, though it is hard to see what other conclusions could be borne out if we endorse, as the objection does, the standard counterfactual view of causal significance.[96] Just as the outcome of a mass election does not depend on your particular vote (except in the rarest of cases, and then in the most marginal sense), the success of collective efforts to avert catastrophic climate change cannot be coherently said to depend on your (or any other) single act of disobedience—or, again, so the objection goes.

The analogy with voting here is not incidental. The same basic conception of causality and rational choice that I am arguing has led scholars away from associating climate disobedience with the end of harm-prevention has also animated a decades-long debate among political psychologists as to why people vote. One of the earliest and most enduringly influential participants in this debate, Anthony Downs, famously argued that there simply is *no* instrumentally rational reason for individuals to vote, given the negligible (essentially zero) chance that their doing so will determine the outcome; in Downs's words: "Each man's ballot is only one drop in a vast sea."[97] Many since Downs have more or less accepted this view,[98] though some have diligently sought to explain why and when voting might nonetheless be rational: i.e. alternative reasons why people would be motivated to participate in mass electoral processes despite the apparent causal insignificance of their doing so. For instance, some have argued that voting affords the voter a kind of "intrinsic satisfaction"; or serves an "expressive" role; or helps to allay anxiety about the guilt one might feel or the bad consequences one might face should one's preferred candidate lose by just one vote.[99] This debate has a direct correlation to academic and popular discussions of the significance of climate activism,[100] in which, again, one finds a nearly ubiquitous tendency to impute

[96] Or, perhaps better said, standard *today*; as Tuck (2008) compellingly shows, the predominance of this view is recent.

[97] Quoted in Tuck 2008: 30.

[98] See e.g. Brennan 2011 for an almost caraciturish example; cf. Maring 2016. Of course, others have challenged Downs's view, perhaps most notably including Derek Parfit (1984). For a more recent discussion, see Barnett 2020.

[99] For an excellent discussion of the literature, see Tuck 2008: ch. 2, esp. pp. 32–4.

[100] Lane (2018) and Tuck (2018) also note this correlation, although not with respect to climate activism in particular, but rather to individual climate action in general.

communicative ends as apparently the only coherent basis for regarding such actions as meaningful and rational.

I claim that we should reject this view. Of course, a full rebuttal would require a much lengthier argument than I can provide here. I nevertheless hope that my remarks in the following sections, provisional as they may be, will lend plausibility to the view that certain individual acts of climate disobedience are, or at least can be, causally significant and so *necessary* for preventing catastrophe, thus qualifying them as instances of HP-CD.

4.4.2 The significance of individual contributions to collective efforts

To begin, an analogy with voting is again helpful.[101] Imagine an election between two candidates, Isabella and Lupe. In this election, there is a determinate[102] threshold of success: if either candidate achieves a majority (50 percent + 1) of the votes tallied, she wins. Suppose that, in the end, 100,000 votes are cast, and Lupe loses badly, having secured just 39,999 votes. In this scenario, Isabella needed only 40,000 votes to win, although she received many more (60,001). Critically, though, until the 40,000-vote threshold was reached, each and every vote that Isabella received was necessary in bringing about her victory. More specifically, adopting Richard Tuck's helpful terminology, 40,000 of the 60,001 votes that Isabella received constituted a causally "efficacious set."[103] The efficacy of this set depends upon each and every vote comprising it, considering that, had Isabella received only 39,999, she would not have won but tied, while any fewer would have been a loss. Thus, *each* of the votes she received up to and including the 40,000th were *necessary*. Of course, the votes in excess of that 40,000 threshold did not constitute part of the efficacious set, and in this sense were *not* necessary for bringing about Isabella's victory. Yet we can still say, following Tuck, that each vote cast for Isabella had a roughly 66.5 percent chance of belonging to the efficacious set.[104] Should the margins have been narrower, the probability would have been even higher; if, for instance, Isabella received 52,000 votes, and Lupe 48,000, then each vote for Isabella would have had about a 96 percent chance of belonging to the efficacious set, and thus of being necessary for producing her victory. The key point here is that, in cases like these, each individual contribution is causally efficacious

[101] What follows leans heavily on Tuck's (2008) outstanding book on free-riding.

[102] Or "absolute"—the terminology varies in the literature. On this, see Parfit 1984: 73–82. For helpful secondary analysis of the relevance of thresholds in these cases, see Klosko 1990; Lane 2018: esp. 162–3.

[103] Others, including Parfit (1984), have similarly emphasized the extent to which actions that form parts of sets can thereby gain causal significance.

[104] Calculated as [Total votes necessary for Isabella to win] / [Total votes Isabella received] (Tuck 2008: 44).

and, in that way, necessary for bringing about the final outcome up to the point at which a determinate threshold is surpassed. As is probably already clear, this view cuts against the common, counterfactual account of causal significance, which in this case would stipulate that only the "pivotal vote" (i.e. the first that tops the threshold) is necessary for deciding the election.[105]

But what does all of this have to do with climate disobedience? I argue that, much like winning an election, preventing climate catastrophe is a goal that can be defined in relation to a warming-limit threshold.[106] A widely agreed-to upper bound for ensuring noncatastrophic outcomes (at least on a global level) is 1.5°C.[107] It is worth emphasizing, though, that many scientists believe that even before reaching that limit, and certainly after, there is a risk of hitting yet-unknown "tipping points," which, as noted in Chapter 2, are themselves "*thresholds* beyond which major changes occur that may be self-reinforcing and are likely ... irreversible over relevant time scales."[108] The most likely and most disconcerting tipping points are the destabilization of polar ice-shelves (which could precipitate sudden and devastating increases in sea level, and, with those, new and extreme weather patterns) and the thawing of arctic permafrost (which could release enormous quantities of methane and thus result in "runaway global warming"[109]). Again, we are not sure what level of warming would be necessary to surpass these thresholds, but we do know that tipping points constitute a credible risk of political catastrophe on a global scale,[110] and that the only viable way to prevent this, at least with existing technologies,[111] is by reaching net–zero emissions as quickly as possible. Given this, 1.5°C is only a rough threshold: one that provides modest (not complete) hedging against extreme risks, which are themselves linked to other, unknown (but nonetheless determinate) thresholds.[112] Some, of course, may object to regarding 1.5°C of warming as the relevant threshold, considering it either too conservative or overly lax (given the catastrophic potential of tipping points, which may be triggered earlier[113]). For my argument to hold, though, it is not necessary to specify an exact threshold, only to assume that one exists, below which politically catastrophic climate change can be avoided, and above which it cannot.[114] We can call this the catastrophic warming threshold, or CWT.

[105] Tuck 2008: 44, 51.
[106] On this, see Lane's (2018: 162–3) helpful discussion.
[107] Steffen, et al. 2018; Armstrong McKay, et al. 2022.
[108] Furman, et al. 2014: 20. See also Wunderling, et al. 2023. On regarding tipping points as catastrophic, see McKinnon 2012: 48ff.; Lenton, et al. 2019.
[109] See Whiteman, et al. 2013; cited in Wagner and Weitzman 2016: 185.
[110] See Chapter 2, section 2.6.1.
[111] On this, see Mittiga 2019c: 474ff.
[112] IPCC 2018b.
[113] Wunderling, et al. 2023.
[114] For an argument along these lines, see Tuck 2008: ch. 3, esp. pp.95ff. Lane (2018: 162–3) makes a similar point about the practical (in)significance of indeterminacy with respect to climate thresholds. For more general discussion of this issue (outside of the climate context), see e.g. Kagan 2011; Ackerly 2018.

As just mentioned, to avoid surpassing the CWT, our most critical task is to reduce greenhouse-gas emissions as quickly and extensively as possible. Just how quickly and extensively is also a matter of debate, but we can specify probabilistic ranges. For instance, according to the most recent IPCC report, if we are to have a greater than 50 percent chance of not surpassing the CWT, we must achieve "reductions of 43% [34–60%] by 2030 and 84% [73–98%] by 2050."[115] These emissions-reduction targets (ERTs) constitute another kind of threshold, one which we must reach if we are to have a better-than-even chance of not surpassing the CWT. Alarmingly, even with the Paris Agreement pledges, we are far from meeting these ERTs, and so currently are running a very high risk of overshooting the CWT.[116] Put simply, if our goal is to prevent catastrophe, we are failing badly.

This is where climate disobedience steps in.[117] Climate disobedients operate with the assumption that the ERTs can be met, and thus that catastrophe can still be prevented. (I will say more about this assumption below, in section 4.5.2.) In attempting to achieve the ERTs, they understand that every reduction (R) of emissions counts. There is of course vagueness here: any particular R could be very large or very small. But we can at least assume that, while (i) no single R will be sufficient, in itself, for meeting the ERTs, (ii) each R will be >0, on the plausible ground that any reduction of emissions, however small it may be, is greater than no reduction at all.[118] It therefore follows that an unknown, but nonetheless determinate, aggregation of *R*s $|\{\mathbf{R_1}, \mathbf{R_2}, \ldots, \mathbf{R_n}\}|$ will be sufficient for meeting the ERTs, and thus keeping below the CWT.

Now, if we assume, as I claim climate disobedients do, that the ERTs can and will be met, then each and every R undertaken up to R_n will have been necessary for bringing about that outcome—i.e. preventing, or at least limiting the risk or scale of, catastrophic climate change. In other words, each and every R, up to R_n, will have constituted part of the "efficacious set." Clearly, it is impossible to identify which action will be R_n, though this is of little consequence, for two reasons. First, because, as with elections, it is not the "pivotal" contribution—i.e. the first in the series to surpass the relevant threshold—that is determinate, as that contribution could not have produced the outcome if all the others in the set ($R_{<n}$) had not already been made. In other words, R_n is not in itself causally sufficient, any more than any $R_{<n}$; its significance owes entirely to it being part of the broader set ($\mathbf{R_1} + \mathbf{R_2} + \ldots \mathbf{R_n}$). Second, and more critically, while we cannot say when R_n

115 IPCC 2022b: 21–25. See also IPCC 2018; IPCC 2018b; and IPCC 2021b.

116 See e.g. Liu and Raftery 2021.

117 For the remainder of the chapter, where I speak of climate disobedience, it should be understood specifically as that in the mold of HP-CD.

118 An exception here would be if a given reduction somehow prevents a greater one. But this seems highly unlikely in most cases, and thus it falls to critics to establish that this would be a significant enough issue to merit discarding the simplifying assumption that every R is >0.

will occur, we can say with virtual certainty that any current *R* will be $R_{<n}$, given that we are still far short of meeting the ERTs. It follows, then, that every *R* undertaken today is *necessary* for preventing catastrophe, at least if we assume that the ERT will be met. (Again, I defend this assumption below.) Notably, this is true even if other technologies, like carbon capture or geoengineering, begin to limit the need for mitigation, given that how much (say) carbon must be sequestered or how many geoengineering interventions must be deployed to prevent catastrophe will depend on the *R*s undertaken up to that point. In effect, such technologies, along with conventional climate abatement, are also *R*s (or *R*-equivalents).

Of course, if disobedients are right in assuming that the ERTs will be met, then at some point R_n will occur and thus every subsequent action will be unnecessary.[119] We should note moreover that those *R*s that follow from specific acts of disobedience may become unnecessary *before* the ERTs are reached if there are strong grounds for thinking that governmental, or at least legally permissible, action will suffice. For regarding disobedience as "necessary" depends not only on its causal efficacy but also on the context in which it is undertaken—which, on most accounts, must be one in which there exists a lack of viable legal means for preventing harm.

This raises an important practical question: how will we know when no further acts of disobedience are necessary (i.e. that a given *R* is $R_{\geq n}$)? Given the irresolvable ambiguities of climate science, responding to this question requires a kind of Aristotelian practical wisdom.[120] That is, given the (probabilistic) vagueness of the relevant thresholds, the complexity of projective modeling, the difficulty of precise measurement, etc., when deliberating over whether any particular *R* is $R_{<n}$ or $R_{>n}$, some degree of phronetic judgment will be necessary.[121] This is part of why it is crucial that HP-CD activists submit themselves to legitimate judicial authority. For we expect precisely this kind of (phronetic) wisdom to preside in the courtroom—if not in the judges then at least in the juries. And so in submitting themselves to the decisions of a court, disobedients are, in effect, exhibiting a distinct form of fidelity to law—in particular, fidelity to the part of law that is bound up with fair and prudent human judgment.

Now, of course, courts can err. To limit this, judges and juries would be well advised to adopt a "precautionary posture."[122] In more concrete terms, truly phronetic judgment in the context of assessing responses to grave risks should: not

[119] To be clear, I am only claiming here that every subsequent action ($R_{>n}$) will be unnecessary *with respect to the end of preventing catastrophe.* Nevertheless, such actions would remain significant insofar as they continue to the broader end of limiting or preventing (noncatastrophic) climate harm (as Lane [2018: 163–4] suggests)—much as votes in excess of a threshold of victory may remain valuable with respect to other, secondary ends, like strengthening the elected candidate's mandate (Guerrero 2010; cf. Lane 2018: 165–6).

[120] See *Nichomachean Ethics*, 1140a25ff.

[121] Tuck (2008: 88) makes something like this point.

[122] See Chapter 2, section 2.5.1.

regard a lack of full certainty as a reason not to act; support action that serves to prevent or minimize loss and damage, especially that which is catastrophic or irreversible; and favor measures that provide greater over lesser protection, subject to reasonable limits. In the simplest terms, prudence in this context would call for erring on the side of caution, which means treating any particular R as $R_{<n}$ until the ERTs have *unequivocally* been met.

4.5 Additional objections

We now have a clearer understanding of how individual acts of civil disobedience could be considered "necessary" for preventing climate catastrophe. But several questions remain regarding: (i) the nature of necessity in this context; (ii) the grounds one has for believing that the ERTs will, in fact, be met; and (iii) the relative efficacy of disobedience compared to other potential forms of climate action.

4.5.1 Substitutability and the nature of necessity

The first issue is essentially one about substitutability. A critic could argue that no particular act of climate disobedience is necessary, given that it could be substituted for any other disobedient act of equal or greater magnitude, or even for other (perhaps legal) emissions-reducing actions, like foregoing plane travel or adopting a vegan diet.

I accept the point that any particular act of disobedience could be substituted for another, and is therefore not necessary in that sense of the word.[123] That being said, the mere possibility of substitutability does not negate necessity in the sense of invariably producing a given outcome, which is the relevant conception here.[124] In other words, even accepting that **A + B + C = D** *and* that **A + B + E = D** does not imply that C or E are unnecessary, but only that neither C nor E are *required* for D. This does not change the fact, however, that if, in addition to A and B, we have either C or E, then we will invariably have D. This is akin to saying that having *either* ten nickels *or* two quarters will invariably entail having fifty cents. Of course, having nickels is not "necessary" for having fifty cents in the strong sense that *nickels and nickels alone* can be aggregated so as to yield fifty cents—two quarters, or for that matter five dimes, serve equally well. Nevertheless, if I have ten nickels, I will invariably have fifty cents, and each of those nickels is *necessary* for producing that sum in the obvious sense that if I lose even one, it will no longer be the case that I have fifty cents.

[123] It could also be substituted for climate actions apart from disobedience. I return to this point in section 4.5.3.

[124] Tuck 2008: 58–9.

The same, I argue, is true of climate disobedience. Certainly, any particular act of climate disobedience, R_{CD}, could be substituted for any other equivalent climate action(s), but it does not follow from this that R_{CD} is unnecessary: for so long as R_{CD} will, *in fact*, form part of the efficacious set, then it will have been (for that reason) necessary for having produced the relevant outcome (in this case, meeting the ERTs and averting catastrophe).

We must note, however, that any particular climate action—that is, any particular *R*—including any given instance of climate disobedience, will be necessary *only if* the ERTs are met. Put another way, each and every *R* will have been necessary if and only if they *actually* form part of an efficacious set, just as each and every vote for candidate Y becomes necessary if and only if Y wins the election. In this sense, the concept of necessity I am using here depends on sufficiency: if a sufficient number of actions are taken, then each will have been necessary up to the point of sufficiency; if sufficiency is *not* achieved, however, then none of the actions will have been necessary. This means that we cannot establish the necessity of a given action as a *matter of fact* until the relevant threshold is passed—i.e. in retrospect.

4.5.2 Radical hope and "presumptive necessity"

What, then, do claims of necessity hang on now, in the context of climate disobedience? I argue that when climate disobedients claim necessity, they are animated by "radical hope." This hope is based on three basic beliefs: (1) that climate catastrophe can still be averted; (2) that while the contribution of individual actions to that end may be imperceptibly small, they are not zero;[125] and (3) that a sufficient number of others *will* act before it is too late, such that everyone's efforts will not have been in vain.[126]

These beliefs are "radical" in various ways. Regarding (1), given that there is no assurance of victory, and many reasons to believe that we will fail in preventing catastrophe, any action undertaken today is shot through with a sense of the tragic. In other words, those who act do so in plain (and often painful) awareness that all of their efforts *could* amount to nothing—and, in the case of disobedience, at potentially great personal cost.[127] For (2), as I have already explained, believing

[125] Here, Brian Barry's (1970: 32) criticism of Olson comes to mind: "If each contribution is literally 'imperceptible,' how can all the contributions together add up to something?" (quoted in Tuck 2008: 66).

[126] Consider Albert Bigelow's remarks after being arrested as the captain of the *Golden Rule* voyage: "I do what I do because it is the way I feel about it. My feeling is not the most important thing in the world—it may be all wrong—but it is the best I have. If everyone else in the world will do above all else what he thinks is right about this most important issue [nuclear disarmament], then it shall be properly met. I hope I am a part of this process" (Thompson 1958).

[127] This sense of tragic is similar to that which Nietzsche associates with Greek heroes (Thiele 1990).

that individual action matters cuts sharply against the prevailing counterfactual view of causal significance. Disobedients must therefore contend with the palling, ever-present question, "*what's the point?*," and still summon the will to carry on. Finally, regarding (3) there is something radical—a leap of faith—in undertaking action with the understanding that its significance ultimately depends on a multitude of unknown others acting in kind, and all without the external incentives of public sanction or censure; quite the opposite!—often in the face of strong external incentives *not* to act. After all, right now, virtually any kind of voluntary, individual climate action carries with it something of sacrifice or risk—be that making uncomfortable lifestyle decisions (like going vegan or not flying) or banding with others to overtly violate the law. Moreover, the infuriatingly slow gait of collective progress can make the siren calls of resignation, despair, and denial that much harder to overcome.

And yet, despite the radicality of hope in this context, I believe we have reason to share it. To see this, it helps to consider the alternatives: flat denial, defeatism, or passivism. The first is untenable for anyone who understands the problem in a basic empirical sense. The second, advocated by those like Roy Scranton, effectively amounts to another form of denialism, at least so long as it remains possible to act and prevent catastrophe.[128] Denial in either of these forms is typically made possible only by the kind of insulating comforts and distance that those with great generational luck, residing in the most privileged parts of the world, can afford.[129] The final alternative, passivism, is incompatible with any plausible conception of moral duty—be that a reasonable religious doctrine, the golden rule, utilitarianism, deontology, or virtue ethics.

Together, these beliefs—this radical hope—endow actions undertaken today with what we might call "presumptive necessity." Perhaps paradoxically, regarding disobedience as (presumptively) necessary may contribute to it being so.[130] For when necessity is formally recognized in the courtroom—i.e. when disobedients plead and are exonerated by appeal to necessity—the effect will be to make it more likely that similar actions are taken in the future, and thus that the ERTs will be met, which will in turn mean that all such actions will, in fact, have been necessary.[131] And this is clearly part of what climate disobedients are seeking: to provoke a proliferation of HP-CD—a cascade of action—great enough to prevent catastrophe.

Still, a critic might insist, it's not enough to point out the unattractiveness of alternatives to radical hope. We need some sense of the mechanisms that could lead to a proliferation of action sufficient for realizing an efficacious set. Otherwise,

[128] Malm (2021: ch. 3) makes this point in a compelling critique of what he calls "climate fatalism."
[129] Norgaard 2011.
[130] Of course, this is no more fantastical than believing that a given monetary currency has value or an item of clothing is fashionable—such beliefs are true only and so long as we believe them to be.
[131] Rausch (2019: 602) suggests something like this.

it is not clear what grounds we have for sharing climate disobedients' radical hope, and thus for regarding their actions as presumptively necessary.

I cannot fully respond to this concern here. But we can briefly consider a few points that may help to assuage the critic. The first is that HP-CD is not happening in a vacuum. Many other forms of climate activism are occurring simultaneously, from letter-writing and large-scale marches to civil and uncivil forms of communicative disobedience. Moreover, there are many mixed cases of disobedience: actions that have both communicative and harm-preventative ends. The Valve Turners, for instance, not only closed pipelines but broadcast themselves doing so on social media. And of course there is discussion of these actions on traditional and social media. This undoubtedly helps to convey the severity of the climate crisis, and provides observers with a sense of the many forms of action available to them for responding to it. Experience suggests that this increased awareness can induce people to participate in climate activism, including HP-CD.

Moreover, with HP-CD cases in particular, it seems that part of the ambition of using the necessity defense to make it less risky for others to take part in disobedience is that this may produce a kind of "bandwagon effect." The idea here is that the more people who act, the more rational it becomes for others to do the same, precisely because the likelihood of collective success increases. This may seem paradoxical. And, interestingly, many political psychologists regard the "bandwagon effect" in electoral scenarios as something of a puzzle: why do people vote when they believe that the candidate or option they are supporting is likely to win (and perhaps by a lot), even without their vote?[132] Part of the answer here, as in the climate case, may be expressive—people want to be part of the winning team, and to be seen to be part of the winning team—but, following Tuck, I think another part of the answer can only be understood if we reverse standard assumptions about instrumental rationality. This means recognizing that the more people there are striving toward a collective end, the more (not the less!) rational it becomes for me to do the same,[133] because the more who act, the more assurance I have that our collective effort will prevail and my contribution will not have been in vain. Precisely when and how this critical mass will be achieved is impossible to determine in advance. Yet history is replete with examples of situations in which a trickle of action suddenly becomes a flood, and outcomes that were previously considered far off, unlikely, or even impossible were realized suddenly and all at once.[134]

[132] Tuck 2008: 61.
[133] Tuck 2008: 60.
[134] As Melissa Lane (2018: 167) puts it: "To treat the overall outcome of human interactions as a static given, against which one must assess the negligibility of one's actions, is a form of rationality that takes no account of the ruptures that mark our political history." On this, see also Kuran 1991; Solnit 2016; Katzenstein and Seybert 2018 (esp. chs 1 and 2).

4.5.3 Opportunity costs

A final objection concerns the relative efficacy of climate disobedience compared to other forms of climate action. Specifically, a critic might accept that climate disobedience *could* contribute to the end of preventing harm, as I have been arguing, but question whether it was the *most* effective way of doing so. Here again one can find an analogous objection in the recent literature on electoral democracy, advanced by those who are skeptical that—for most citizens in most cases—voting is the all-things-considered best way of using scarce time and resources.[135] According to this objection, it would seem misguided to undertake climate disobedience, at least if doing so were to come at the cost of other, more impactful actions.

There are many ways we might respond to this objection. As a first point, it is important to note that I have nowhere claimed that disobedience is the best all-things-considered form of climate action. It may well be the case that other modes of intervention are more effective, although that is not obviously so. After all, the means by which ordinary citizens can affect the political process are typically limited, and most—if not all—are similarly "imperceptible" in effect, at least when taken in isolation. One might vote for good climate candidates or policies, but that would be just one vote among many; one might seek office oneself, but (even if successful) that would amount to just one (limited) office among many (even the president in an enormous state like the US is constrained); one might pursue public outreach—say, via op-eds, radio interviews, or social media posts—but this would still amount to just one voice in a broader cacophony. This is all to say that there is no obvious reason for assuming that disobedience is or will be *necessarily* less effective than other forms of individual action. In any case, this is the type of empirical question that is impossible to settle in the abstract. The only claim I have tried to defend here is that it is at least reasonable to consider civil disobedience as *one* potentially effective way of contributing to the end of preventing climate catastrophe—but, again, not the only one and not always or necessarily the best.

It is, however, worth reflecting on the fact that many only turn to disobedience once they feel they have no other viable options. Relevant here are the remarks of Abby Brockway, a climate disobedient who, in 2014, along with other members of the "Delta 5," blocked trains shipping coal and gas through Washington state. In an interview with a local news station just moments before being arrested, Brockway said: "We're ready to take more steps to get this done, because we're running out of time ... And so, this is just a second step. I don't know what the third step is."[136] She later explained that step one was "writing letters, speaking at hearings, participating in legal protests,"[137] with the aim of convincing those in power to take action,

[135] See e.g. Brennan 2011.
[136] Goodman 2016.
[137] Mazza 2016.

which ultimately failed to produce change. Step two, for her, was HP-CD: i.e. the train blockade, which aimed at impeding a significant source of climate harm (oil and coal exports), and which was undertaken in accordance with the basic standards of civility (non-violence, publicity-in-action, willing acceptance of arrest). While Brockway did not specify what the third step was, the implication was clear: should *civil* attempts prove insufficient for preventing climate harm—just as law-abiding, communicative protest action already had—then citizens like her might rightfully begin to avail themselves of more extreme forms of resistance.[138] The point here is simply that most people who are concerned with preventing climate catastrophe do not *start* with disobedience; rather, they end up there after finding other means for effecting change fruitless.

As a final point, it is helpful to keep in mind that participating in disobedience does not exclude one from undertaking other forms of climate action. Quite the opposite. Some disobedients, like Ken Ward (of the Lobster Boat and Valve Turner cases), are active in environmental advocacy and lobbying groups (in his case, Greenpeace). Some, like those involved in the Scientist Rebellion, have dedicated their careers to careful empirical analysis of the climate problem and articulating potential solutions to it. And many, if not most, disobedients maintain strict personal codes of behavior, including avoiding plane travel, adhering to a plant-based diet, composting, voting for green candidates and policies, and so on. Given the scale of the crisis and what's at stake, this all-of-the-above strategy clearly seems warranted, if not morally obligatory. In any case, the essential point here is that there is no reason to believe that disobedience must come at the cost of other, potentially more effective contributions to preventing catastrophe.

4.6. Conclusion

In this chapter, I have defended two basic claims. First, that prevailing theories of civil disobedience fall short insofar as they fail to acknowledge that some actions aim foremost at the prevention of harm—an end that cannot be reduced to, and will not always be paired with, the communicative aspirations normally associated with civil disobedience. Harm-preventative civil disobedience (HP-CD) shares many of the basic features of communicative civil disobedience (C-CD), including a commitment to nonviolence, nonevasiveness, and publicity-in-action. Yet the difference in ends entails a difference in how each instrumentalizes the penal process: whereas C-CD activists often exploit punishment as a means of (e.g.)

[138] This same basic point has been expressed innumerable times by a range of climate disobedients. To take just one other example, consider Tiana Jacout's remarks after participating in an Extinction Rebellion protest that blocked many high-volume roadways: "It is not a step we take lightly," however "[w]e have tried marching, and lobbying, and signing petitions. Nothing has brought about the change that is needed" (BBC News 2018a).

eliciting sympathy or signaling fidelity to law, many HP-CD cases make innovative use of the "necessity defense" as a means of reclaiming public authority over harmful actors and practices. Of course, the need for such action only arises when governments fail to protect against such threats. In this sense, HP-CD is a response to illegitimacy (in the foundational sense of the word[139]) and is justified accordingly.

The chapter's second basic claim was that, among the most politically and legally consequential cases of climate disobedience, at least in the present moment, are those that fit the HP-CD mold. Their significance, however, ultimately depends on their ability to employ the necessity defense successfully. And this suggests a potent objection: that it is highly implausible, if not altogether incoherent, to believe that any single act of civil disobedience could prevent climate harm—let alone the very great harm of climate catastrophe; individual acts are simply too limited in scale, their impact imperceptible. In responding, I drew on a parallel debate over the significance of individual votes in mass elections and argued that, granting certain plausible assumptions, there are strong grounds for regarding individual acts of climate disobedience as (at least presumptively) necessary for preventing catastrophe—however "imperceptible" their effects may be when taken alone.

There are still many questions left to be answered. Perhaps above all are those concerning the relationship between HP-CD and more extreme forms of political action. One possibility, suggested earlier, is that a polity in which ordinary citizens feel compelled to break the law in order to prevent serious harms—especially harms that reasonably *could be* avoided by governmental action but simply are not—would seem to be one in the midst of a serious legitimacy crisis. To the extent that a government fails to preserve basic safety, citizens might rightly question its authority and seek to reclaim some portion of it for themselves, as I have argued some climate disobedients are now attempting to do. If this is right, then perhaps HP-CD can be understood as a prelude to more radical interventions, like rebellion or revolution.

Something like this is suggested in the speeches and publications of various climate disobedient groups. The Climate Disobedience Center, for instance, presents a lengthy defense of "the right to resistance" on its website, which, among other things, argues that:

> Popular consent is the predicate of democratic governance and may be revoked when the institutions of government have lost their allegiance to the public ... The Universal Declaration of Human Rights asserts that the people "have recourse, as a last resort, to rebellion against tyranny and oppression" ... This resistance is

[139] See Chapter 3.

justified by the extreme and existential threat of climate change to all human and non-human life.[140]

The Extinction Rebellion UK's "Declaration of Rebellion" includes similar language:

> When Government and the law fail to provide any assurance of adequate protection, as well as security for its people's well-being and the nation's future, it becomes the right of its citizens to seek redress in order to restore dutiful democracy and to secure the solutions needed to avert catastrophe and protect the future ... We hereby declare the bonds of the social contract to be null and void, which the government has rendered invalid by its continuing failure to act appropriately. We call upon every principled and peaceful citizen to rise with us.[141]

Of course, as the climate crisis deepens and future prognostications grow ever more dire, we can only expect such proclamations to become even more radical and widespread. Already calls for bolder action are being met with considerable (if cautious) interest (as with Andreas Malm's provocatively titled *How to Blow Up a Pipeline*[142]). Increasingly commonplace, too, are vivid imaginings of overtly violent interventions, like the oil executive assassinations and private jet bombings described in Kim Stanley Robinson's semi-dystopian climate novel, *Ministry for the Future*.[143]

This speaks to Hannah Arendt's claim that civil disobedience springs from the same source as revolution. A "disintegration of political systems precedes revolutions," she argues, and "the telling symptom of disintegration is a progressive erosion of governmental authority ... caused by the government's inability to function properly, from which spring the citizens' doubts about its legitimacy."[144] These doubts, left unassuaged, can blossom into crises capable of transforming "voluntary association into civil disobedience and ... dissent into resistance."[145] To avoid this disintegration of legitimacy and peace, governments must urgently follow the lead of their civil disobedients and begin asserting real authority over the agents

140 Climate Defense Project 2018.

141 Extinction Rebellion 2022.

142 For critical responses to Malm, see Armstrong 2022; Battistoni 2022; Riofrancos 2022.

143 Another example of this in popular media is season 5, episode 3 ("A Rickconvenient Mort") of the popular television series *Rick and Morty*, which depicts a frustrated eco-superhero resorting to slashing tires on a gridlocked highway, killing apathetic coal miners, and setting fire to a corrupt, fossil-friendly Congressperson's house.

144 Arendt 1970: 78. Di Paola and Jamieson (2018: 424) also make something like this point: "Unable to find consistent responses to challenges to their own legitimacy in the Anthropocene, liberal democracies may be in danger of warranting revolutions against themselves."

145 Arendt 1970: 105.

and practices that are imperiling the planet and its many vulnerable inhabitants. If they fail to do so, they risk violence, destruction, and collapse. This is the *political* danger of climate catastrophe: the Ozymandian fate of failed power asserted on forgotten works slowly being swallowed by desert sands.[146]

[146] I refer here to Percy Shelley's poem, "Ozymandias":

> I met a traveller from an antique land,
> Who said—"Two vast and trunkless legs of stone
> Stand in the desert ... Near them, on the sand,
> Half sunk a shattered visage lies, whose frown,
> And wrinkled lip, and sneer of cold command,
> Tell that its sculptor well those passions read
> Which yet survive, stamped on these lifeless things,
> The hand that mocked them, and the heart that fed;
> And on the pedestal, these words appear:
> My name is Ozymandias, King of Kings;
> Look on my Works, ye Mighty, and despair!
> Nothing beside remains. Round the decay
> Of that colossal Wreck, boundless and bare
> The lone and level sands stretch far away.

5
Exiting Eden

What is the late November doing
With the disturbance of the spring
And creatures of the summer heat,
And snowdrops writhing under feet
And hollyhocks that aim too high
Red into grey and tumble down
Late roses filled with early snow?
Thunder rolled by the rolling stars
Simulates triumphal cars
Deployed in constellated wars
Scorpion fights against the Sun
Until the Sun and Moon go down
Comets weep and Leonids fly
Hunt the heavens and the plains
Whirled in a vortex that shall bring
The world to that destructive fire
Which burns before the ice-cap reigns.

T. S. Eliot, from "East Coker"

In *Works and Days*, Hesiod claims that in the first, "golden" era of the earth, humans (or some proto-human race) "lived like gods without sorrow of heart, remote and free from toil and grief ... for the fertile earth unforced bore them fruit abundantly and without stint."[1] In his treatise on agriculture, Virgil similarly describes a time, when:

Fields knew no taming hand of husbandmen
To mark the plain or mete with boundary-line—
Even this was impious; for the common stock
They gathered, and the earth of her own will
All things more freely, no man bidding, bore.[2]

The idea of an early age of plenty likewise appears in Genesis: "Out of the ground the LORD God made trees spring from the ground, all pleasant to look at and good

[1] Hesiod: Works and Days. 114–21.
[2] Virgil: Georgics. Bk. 1, 125–8.

Climate Change as Political Catastrophe. Ross Mittiga, Oxford University Press.
© Ross Mittiga (2024). DOI: 10.1093/oso/9780192868879.003.0005

for food."[3] Indeed, the very name "Eden" is suggestive of natural abundance—its Aramaic root signifying a place that is "fruitful, plentiful" and "well-watered."[4] Genesis shares, too, Hesiod's and Virgil's suggestion that agriculture marked the point of our descent from this original, happy state of abundance. After Adam and Eve eat the only forbidden fruit in the garden, agrarian toil features centrally in the package of punishments God metes out: "Because you have ... eaten from the tree which I forbade you, accursed shall be the ground on your account. With labor you shall win your food from it, all the days of your life. It will grow thorns and thistles for you ... You shall gain your bread by the sweat of your brow."[5]

There is a parallel between these mythical accounts and recent archaeological work on the Neolithic Revolution, which similarly suggests that our earliest human ancestors enjoyed such great natural abundance that agriculture was unnecessary to satisfy their basic needs. After all, anatomically modern human beings (*homo sapiens*) predated agriculture by about 300,000 years.[6] Many scholars also believe that the eventual uptake of agriculture made life for most in that time considerably more nasty, brutish, and short. John Lanchester, for instance, writes:

> The fossil record shows that life for [the earliest] agriculturalists was harder than it had been for hunter-gatherers. Their bones show evidence of dietary stress: they were shorter, they were sicker, their mortality rates were higher. Living in close proximity to domesticated animals led to diseases that crossed the species barrier, wreaking havoc in the densely settled communities.[7]

The turn to agriculture also gave rise to much heavier (and often coerced) labor than had existed in traditional hunter-gatherer societies. We can still see this today. A 1966 study of the Ju/'hoansi—a still existing, if increasingly beleaguered, hunter-gatherer tribe in southern Africa (mostly in Botswana, Namibia, and Angola)—found that the average tribe member worked about 17 hours a week to procure food and devoted another 19 hours to domestic chores, while maintaining an average daily caloric intake of 2,300. Compare that to the contemporary US worker, who each week spends 40 hours at work (putting "food on the table") and nearly 36 more on domestic chores, while consuming roughly the same number of daily calories.[8] The cost in extra work and ill health has led Jared Diamond to call the Neolithic Revolution "the worst mistake in human history."[9]

What would lead us to make such a mistake? The probable answer is that it was less a choice than a matter of circumstance: approximately 12,000 years

[3] Coogan, et al. 2010: Genesis, 2:9.
[4] Cohen 2011: 229.
[5] Coogan, et al. 2010: Genesis, 3:17–19.
[6] Callaway 2017.
[7] Lanchester 2017.
[8] Lanchester 2017, citing Suzman 2017.
[9] Quoted in Lanchester 2017.

ago, a changing climate abruptly concluded the conditions of natural abundance until then enjoyed throughout much of the world, forcing many of our ancestors into what one commentator has called "late-Neolithic multispecies resettlement camps."[10] This climatic change occurred when massive ice sheets fractured and fell into the ocean, shutting down the Atlantic thermohaline circuit and causing global temperatures to plunge between 2 and 6°C. In a matter of years, glaciers stretched down to southern Portugal, trapping enough water to precipitate a severe and protracted drought throughout the Northern Hemisphere.[11] This dry, cold period—the "Younger Dryas"—endured for about 1,200 years,[12] during which time previously stable food supplies became far less dependable. Hence, our ancestors likely took up agriculture as a matter of necessity: in order to feed themselves in a new age of climatic change and material scarcity.[13]

The agrarian societies that formed were marked by much more complex social, economic, and political organization. Farmers' crops became the basis of the first tax collection systems (cereal crops were favored for being easy to assess and weigh), and their land the first real fiefdoms requiring protection. This spurred the formation of distinct professional classes (farmers, soldiers, priests, rulers) and a corresponding societal division of labor, along with some of the earliest sciences (especially agricultural engineering and meteorology).[14]

Eventually, however, the Younger Dryas ended, giving way to our current geological epoch—the Holocene. This epoch has been characterized by remarkable climatic stability and relative warmth—evidently the most stable climatic interval in at least 650,000 years.[15] The favorable conditions of the Holocene certainly could have supported a return to hunting and gathering. Yet—perhaps because our post-agricultural populations had grown too large, or because social-institutional inertia had already become too great—once the glaciers of the Younger Dryas receded, humans largely persisted in their new agricultural societies.

Our species nevertheless began to flourish in different ways during this second, fecund period. Virtually every work of art, every scientific discovery, the whole of our written history, and certainly every major civilization is an artifact of the Holocene—i.e. everything from Babylon to Buddha, Beethoven, blue jeans, the Japanese Shogunate, Sappho, Socrates, constitutional democracy, the Catholic Church, the Kama Sutra, the Olmec, Marie Curie, the airplane, Al-Farabi, pensions, penicillin, Nazis, nukes, and telenovelas. What makes this so remarkable is that the Holocene has been such a short episode in our species' history, and a mere

[10] Scott 2017.
[11] Gardiner 2004: 562–3.
[12] Scranton 2015: 32.
[13] Anderson, et al. 2011. Interestingly, Gowdy (2020) argues that climate change may ultimately necessitate a reversion to hunting and gathering.
[14] Scott 2017. See also Scranton 2015: 32.
[15] Scranton 2015: 38, citing Archer 2016: 130; see also Rockström, et al. 2009.

blip in the geographical record. As one commentator puts it: "If human existence on Earth were a day, our approximately five millennia of recorded history would take up the last half hour before midnight."[16]

Nevertheless, in providing amply for our needs while supporting an astounding array of cultural and biological diversity, the Holocene has constituted its own kind of Golden Age, even if the rigid social and ideological structures we have imposed on this world often obscure the fact. Without immediate and drastic action, however, we are soon to be evicted from this second Eden—from humanity's "climate niche"[17]—and plunged into a new epoch marked by severe scarcity, conflict, and collapse.

Perhaps this second "fall" became inevitable the moment James Watt's steam engine first began operating in 1781. Fueled by England's vast supply of coal, the steam engine powered the Industrial Revolution by providing a portable, efficient, and continuous source of energy.[18] From that time on, industrial development has been tied to the combustion of fossil fuels. Two centuries later, fossil-fuel industrialization is globally ubiquitous, providing for an unprecedented accumulation of wealth *and* atmospheric greenhouse gases (GHGs).

The climatic consequences of this are now everywhere around us. Going as far back as we can accurately measure (about 800,000 years), the atmospheric concentration of carbon dioxide (CO_2)—one of the most prevalent GHGs—never exceeded 300 parts per million (the pre-industrial average was about 290 ppm). Today, however, it is well above 400 ppm, and steadily rising.[19] At the same time, the atmospheric concentration of methane (CH_4)—an even more potent, if more short-lived, GHG, deriving chiefly from animal agriculture and shale-gas production (i.e. fracking)—has nearly tripled since the Industrial Revolution, and far exceeds any other point in the 800,000-year ice-core record.[20] It is likely that the last time earth's climate had a similarly dense concentration of CO_2 and CH_4 was about 3 million years ago, in the Mid-Pliocene era.[21] During that time, average global temperatures were about 2–3°C hotter and sea levels about 15–25 meters (50–82 feet) higher.[22]

Should such changes come about again, they would be disastrous for the human civilizations we have built. Yet they are far milder than what actually may come to pass. On our present trajectory, by the end of the century we are likely to reach GHG concentration-levels not seen since the Paleocene-Eocene Thermal

[16] Scranton 2015: 33.

[17] Xu, et al. 2020. This basic concern is shared by many working on climate in the emerging existential-risk community; see e.g. Kareiva and Carranza 2018; Xu, et al. 2020; Beard, et al. 2021; Bradshaw, et al. 2021; Huggel, et al. 2022; Kemp, et al. 2022; Ripple, et al. 2022; Steel, et al. 2022.

[18] Scranton 2015: 34.

[19] Wagner and Weitzman 2016: 49–51.

[20] Lüthi, et al. 2008.

[21] Scranton 2015: 34.

[22] IPCC 2007: ch. 6.3.2.

Maximum (PETM), which occurred around 55 million years ago. During the PETM, enormous amounts of carbon and methane were released into the atmosphere over a period of 10,000 to 20,000 years—brief on a geological scale, but still orders of magnitude lengthier than the few hundred years it will have taken us to reach similar levels.[23] With that altered atmosphere, temperatures rose rapidly, eventually peaking at about 5–8°C above pre-PETM levels.[24] This, in turn, led to sea levels nearly 100 meters (300 feet) higher than today's and an arctic circle populated by palm trees and crocodiles.[25] Almost a million years passed before temperatures returned to prior levels.

Today GHGs are entering the atmosphere at a rate *ten times faster* than occurred during the PETM.[26] Without immediate and extensive action, we now face a roughly 10 percent chance of temperatures increasing 6°C *by the end of the century*.[27] In other words, we may force a PETM-like shift not over tens of thousands of years, but in a matter of decades. The fallout from a change so momentous and rapid is unfathomable, in part because it would be geologically unprecedented. But in the short term we could certainly expect to face far more frequent and violent weather events, mass displacement of people and animals, biodiversity collapse (exacerbating the "sixth mass extinction event" that is already underway), and large-scale ecological and agricultural disruptions. And this would just be the result of linear increases in temperature. Even more dramatic and abrupt changes (caused by "tipping points") will likely occur along the way, including the rapid disintegration of the Greenlandic and West Antarctic Ice Sheets, the massive release of methane currently embedded in frozen lakes and land, the desertification of the Amazon, and much more. The melting of the ice sheets, alone, which may already be underway, would entail enormous amounts of cold water flushing into the Atlantic Ocean; this, in turn, could stall the Atlantic thermohaline circuit, potentially triggering a repeat of the Younger Dryas period, in which, again, most of continental Europe became entombed beneath thick glaciers in about ten years.[28]

Changes like these would cause unimaginable suffering and political strain. As fresh water disappears and crops wither, scarcity will create desperate situations. Many people will simply be unable to meet their basic needs; and many of those who can will be able to do so only by denying others the ability to do the same. In such extreme, zero-sum conditions, realizing or sustaining relations of justice becomes difficult if not impossible. Ordinary morality becomes supererogation.

[23] Why this occurred is still unclear, though increasingly scientists think it had to do with large-scale volcanic eruptions (Cui, et al. 2011; Painting 2011).

[24] IPCC 2013: ch. 5.3, esp. p. 399; McInerney and Wing 2011.

[25] Scranton 2015: 35.

[26] Penman, et al. 2014.

[27] Wagner and Weitzman 2016: 53.

[28] See Chapters 1 and 2 for further discussion. On the likelihood of this occurring, see Ditlevsen and Ditlevsen 2023.

And once this occurs—that is, once justice no longer meaningfully applies—even the most robust social and political orders will inevitably fall apart. This is what makes extreme and enduring scarcity *politically catastrophic.*

Despite the great extent of our learning and technological advances over the last 12,000 years, the situation we are facing is far more challenging than that which our ancestors had to confront in the beginning of the Younger Dryas. Whereas the climatic changes that lead to the Neolithic Revolution made justice and political order *necessary* (by transforming a state of natural material abundance into one in which larger-scale social cooperation became necessary to satisfy everyone's basic needs), the climatic changes we are causing here and now threaten to make justice and political stability *impossible*. We are undermining the very foundations of life. Whether or not all of us today realize it, then, we live in a time of great exigency. Our window to prevent politically catastrophic climate change is quickly closing, and if we fail to act in time, it is not clear that humanity, or at least human civilization, will be able to endure what follows.

Of course, well before total collapse, we are likely to witness the rise of potent new forms of totalitarian control, rationing and deprivation, desperate mass migrations (and the fascistic reactionism that so often accompanies them), neocolonial annexations and appropriations, domestic disturbances, interstate conflict, and large-scale environmental desolation. These are the politically catastrophic consequences of extreme scarcity. Should we fail to prevent such dismal outcomes while we still have time to do so, we will have to reckon with the moral anguish of abandoning others—perhaps including our children and grandchildren—to unquenchable thirst, insatiable hunger, violence, and premature death. Such devastation would be clearly condemnable if it were deliberately brought about; are those who allow it to transpire unnecessarily, out of apathy or aversion to reasonable sacrifice, not also culpable?

If we are to have any chance of walking back from this precipice of despair, we must begin now—despite the costs involved and despite the fact that success is in no way guaranteed. In undertaking such action, we may be aided by the radical hope that the world *will* wake up, that others *will* join us in solidarity, and that through our combined efforts the most catastrophic effects of climate change *will* be averted. But this is not a hope that can be nurtured by blithely looking away; by waiting for a technological silver bullet or the work of unknown others. Rather, we must each of us acknowledge the very real possibility of catastrophe, and resolve to do what we can to prevent it.

Of course, we might wonder how to even begin when, as Greta Thunberg once asserted, "the politics and solutions needed are still nowhere in sight."[29] Certainly, the climate crisis has made this much clear: contemporary political thinking and practice has delivered us to this point, and is apparently inadequate for walking us

[29] NPR Staff 2019.

back. Preventing catastrophe, then, requires us to rethink, and perhaps radically revise, some of our most basic political assumptions, along with the norms, institutions, and processes they legitimate and sustain. The central ambition of this manuscript has been to contribute to that end.

***

The inquiries pursued here are, however, only the beginning of a political theory more fully calibrated to the reality of the climate crisis; many questions remain to be answered. Consider, for instance, recent reports that, if we are to prevent catastrophic climate change, we must keep the vast majority of known fossil fuel reserves in the ground. Does justice require us to compensate the oil barons and energy executives who own those reserves?[30] Or, rather, does it require us to punish them for continuing to invest in dangerous assets long after their potential for harm became clear?[31]

Addressing these questions may also require revisiting debates over personal responsibility. Ultimately, we all contribute to climate change, if often in mundane ways by, for example, heating or cooling our homes, consuming animal products, or traveling. Determining whether or not people have a moral duty to stop (or at least limit) such behaviors depends considerably on how we understand the nature of the harms being done and the intentions of the agents in question. These points remain contentious. But invoking moral responsibility may be unnecessary to ground individual obligations. Rather, if climate change presents an existential threat to political stability and just social cooperation, as I have argued, then maybe our responsibility to address it is better understood on the model of conscription during a just war or disaster-relief effort—i.e. as a shared, positive duty, which may be legitimately enforced by governments, and which is valid independent of culpability for past actions. Although some work has already been done on this question, future research would benefit from reflecting more deeply on how the prospect of catastrophe shifts standard assumptions about what individuals ought to (be made to) do, with or without reference to their responsibility for what has already been done.

Moreover, work is urgently needed concerning how we ought to navigate conflicts between liberal-democratic norms and institutions, on the one hand, and the often inescapably technocratic and unpopular nature of many climate policies, on the other. Especially relevant here are practical questions, including, for instance, how states might defuse or at least mitigate popular backlash domestically, and how they might effectively compel recalcitrant actors internationally.[32]

[30] John Broome (2020) suggested something along these lines in a recent talk.

[31] Should we, for instance, charge them with "crimes against humanity," as Catriona McKinnon (2017) has argued?

[32] Bradshaw, et al. 2021: 5.

Of course, popular resistance cuts both ways—both against climate action and for it. And we are now experiencing the emergence of a strong, visible, and committed climate movement—one capable of holding increasingly large and international marches, and undertaking ever more disruptive disobedience campaigns. As the crisis deepens, we can expect such action to continue to grow in scale and frequency, and perhaps take on more extreme forms. After all, if we take seriously the increasingly common comparison of climate change to war, we may naturally begin to wonder: When will censure give way to sanction, disruption to sabotage, civil disobedience to rebellion or revolution?

Finally, and most abstractly, political theorists and climate ethicists must begin to reassess systematically the origins and effects of anthropocentrism in contemporary thought and practice. This would help clarify the duties we owe to nonhuman animals, who, in many ways, face more harm and risk from climate change than we do. Such research may also provide further insight into the broad range of environmental concerns that do not always, directly, or obviously impact human beings, like preserving wilderness areas and preventing biodiversity loss, all of which are affected greatly by climate change.

Most of these questions will remain relevant regardless of the path politics takes in the years ahead. Indeed, even if states were to suddenly undertake decisive measures against climate change, as they once did to avert the ozone crisis, our current predicament has already made clear the fragility of the natural environment, and thus too of the political edifices we have constructed atop it. Our continued disregard of ecological limits not only imperils our ability to live freely, democratically, peacefully, and justly today, but also the very conditions that make freedom, democracy, peace, and justice possible for the foreseeable future. Climate change confronts us now as an almost divinely destructive force, a shard of hell; only our most serious efforts can save our corner of Eden from the coming inferno.

APPENDIX

Further Notes on Climate Change and Scarcity

Extreme weather events, like long droughts and massive cyclones, are just two ways that climate change threatens extreme scarcity, and thus political catastrophe. Consider seafood, a dietary staple for a fifth of humanity.[1] Ocean warming, along with the straightforward contamination (e.g. mercury) caused by the same pollutants that are heating the planet, increase the likelihood that marine life—and those who consume it—will become sick.[2] Moreover, ocean warming is driving marine species toward the poles. This could have dramatic effects on many domestic economies; according to the IPCC, with just 2°C of warming, fish catches in the tropics are likely to "drop [by] 40–60%."[3] This is a major threat to seafood-dependent nations, some of which are among the hungriest in the world.[4]

In addition to the problems caused by warming, increased carbon dioxide (CO_2) levels are also making the earth's oceans significantly more acidic,[5] which again has vast implications for marine life.[6] Ocean acidification prevents animals like mollusks and crabs from forming hard shells (as calcium carbonate dissolves in acidic water).[7] It decreases birth sizes in marine life and impairs sensory perception in many species like salmon, potentially permanently depressing population sizes or simply causing extinction.[8] Ocean acidification also creates or exacerbates "dead zones," which can devastate whole fisheries—an effect well underway in the Gulf of Mexico and around Namibia.[9] Perhaps most disconcertingly, ocean acidification is causing massive coral reef die-offs—i.e. "coral bleaching." In a survey conducted in 2022, a "devastating" 91 percent of the coral in the Great Barrier Reef had been affected by bleaching.[10] This is a significant problem, given that live coral reefs provide the breeding grounds for many species—including ones that humans rely on for sustenance.[11] (Food shortages owing to coral bleaching have already been felt in Kenya, after nearly half of the nearby reefs died in 1998.[12]) Alarmingly, according to the IPCC, "over 40% of coral reefs and 60% of coastal mangroves in the Asia-Pacific region have already been lost, and approximately 80% of the region's coral reefs are currently at risk."[13]

[1] Morello 2012.
[2] Marques, et al. 2010.
[3] IPCC 2014a: Part A, p. 414.
[4] Boyce, et al. 2022.
[5] Currently, the ocean is about 30 percent more acidic than it was before the Industrial Revolution (Morello 2012).
[6] According to Boyce, et al. (2022), "[u]nder high emissions (SSP5–8.5), almost 90% of ~25000 species are at high or critical risk."
[7] Casey 2015.
[8] Asch 2015; Casey 2015.
[9] Minogue 2014; Wallace-Wells 2017.
[10] Morton and Cox 2022.
[11] See e.g. Storlazzi, et al. 2018; Duvat, et al. 2021.
[12] Morello 2012.
[13] IPCC 2022a: ch. 10, p. 1482.

And of course all of these climate-related effects are happening alongside, and thus amplifying, the problem of overfishing, which is also exerting enormous pressure on sea-life. An illustrative example is the near complete collapse of the arctic snow-crab population. Between 2018 and 2021, an estimated 90 percent of these animals "disappeared," owing to a combination of relentless overharvesting and extreme warming in the Bering Sea. Apart from the tremendous toll on the crabs themselves—many of whom likely starved to death—this has had a serious economic impact on small, independent crabbers.[14]

Land-based agriculture is of course similarly vulnerable to a changing climate.[15] One major area of risk involves sea-level rise. Glacier melt and thermal expansion are both elevating sea levels at an accelerating rate. According to one National Oceanic and Atmospheric Administration (NOAA) report, without serious efforts to reduce emissions, we could see as much as 2.5 meters (roughly 8 feet) of global sea-level rise by 2100.[16] This would critically imperil rice harvests. As Lester Brown explains, just 0.91 m (3 feet) would be enough to "inundate half the rice-land in Bangladesh, home to 160 million people" and "devastate the rice harvest in the Mekong Delta, which produces more than half the rice in Viet Nam, the world's number two rice exporter." A 1.88 m (6-foot) sea-level rise would threaten "[e]very rice-growing river delta in Asia."[17] And a 2.5 m (again, roughly 8-foot) rise would likely precipitate unprecedented levels of starvation, particularly in Asia.

Crops farther inland are hardly less vulnerable to climate change. For every degree (centigrade) of warming, scientists expect crop yields to decline by about 10 percent (some estimates range up to 17 percent).[18] And experience bears this out. Having recently exceeded approximately 1°C of warming, the world has already begun to witness serious crop failures in many regions. The mechanisms for this are manifold:

(a) Even short exposure to very high temperatures—say, through heatwaves, which are becoming more common—critically damages plant tissue and depresses harvest yields.
(b) Higher temperatures mean more rapid plant development, resulting in smaller plants and thus lower total yields.
(c) Higher temperatures cause higher rates of water evaporation in soils, killing plants in the short term and contributing to the aridification of arable land in the long term.[19]

Each one of these effects is potentially catastrophic. Consider, for instance, that the US currently produces about 41 percent of the global supply of corn and 38 percent of soybeans—staples across the world. Corn and soy, however, are both highly sensitive to temperature. For this reason, some estimate that "average yields" may "decrease by 30–46% before the end of the century under the slowest warming scenario and decrease by 63–82% under the most rapid warming scenario."[20] In either case, the result of such depressed harvests would be a food crisis of unimaginable scale and intensity—as prices skyrocket, millions (if not billions) of people would be left hungry.

It may be tempting to think that technology will save us from these shocks to the food system, as it has so many times before. After all, predictions about the limits of agricultural

[14] Diaz 2022.
[15] See e.g. Mehrabi 2020; Kornhuber, et al. 2023; Richards, et al. 2023.
[16] Sweet, et al. 2017.
[17] Brown 2009: 7.
[18] Battisti and Naylor 2009; Wallace-Wells 2017.
[19] Biello 2009; Schlenker and Roberts 2009; National Research Council 2011: 159ff.
[20] Goldenberg 2014.

expansion have been made and discredited many times since Thomas Malthus's first ominous declaration. But there are reasons to be pessimistic about the potential for Promethean solutions to our emerging predicament.

Consider the agricultural technologies ushered in since the 1940s, which collectively comprise the "Green Revolution": e.g. the breeding of disease-resistant, high-yield crops; the use of potassium fertilizer and chemical pesticides; the introduction of intensive plowing machines; and the adoption of new, highly efficient irrigation techniques. These technologies were meant to feed a rapidly growing human population and, in important ways, succeeded at this. Yet, as Les Thiele notes, this revolution also had a number of unintended consequences, "many of dire significance."[21] Most critically, contemporary agricultural practices have (1) severely depleted virtually every aquifer in the world;[22] (2) eroded topsoil—some believe we have lost nearly a third of all arable land since 1960;[23] (3) expanded deserts (especially in Asia and South America);[24] and (4) seriously contaminated waterways and deltas via pesticide and excrement runoff, creating or exacerbating oceanic dead zones.[25] In short, for all its success, the Green Revolution has created immense new problems. It increasingly seems that its suite of technological innovations may have only temporarily saved us from the Malthusian problem—and at the cost of rapidly exhausting much of the earth's natural resources.[26]

Even if we could find ways of indefinitely sustaining the productivity gains afforded by the Green Revolution without further (and perhaps irreparably) damaging our arable land and water supplies, we must confront another dilemma: the world population (and thus food demand) is greatly increasing precisely at the moment in which we are reaching the limits of photosynthetic efficiency.

To explain: around the beginning of the Industrial Revolution in 1800, there were roughly 1 billion people on the planet. By 1927 the human population had doubled to 2 billion. We hit 3 billion in 1960, 4 billion in 1974, 5 billion in 1987, 6 billion in 1999, 7 billion in 2011, and are projected to top 9.7 billion by 2050 and 10.4 billion by 2100.[27] This meteoric growth rate over the last century can, in part, be explained by the same unsustainable agricultural practices that imperil us today; by ensuring that many more people could be fed in the short term, the Green Revolution paradoxically gave rise to a situation in which there would be many more people to feed over the long term.[28] In fact, according to the UN Food and Agriculture Organization (FAO), "feeding a world population of 9.1 billion people in 2050 would require raising overall food production by some 70 percent between 2005/07 and 2050," which entails that "[p]roduction in the developing countries would need

[21] Thiele 2013: 104.

[22] Brown 2009: 14, 32.

[23] Brown 2009: 4ff., 32ff.; Thiele 2013: 104. According to Volkert Engelsman, an activist with the International Federation of Organic Agriculture Movements, "We are losing 30 soccer fields of soil every minute, mostly due to intensive farming" (quoted in Arsenault 2014). At this rate, by 2050, we will have only one-quarter of the arable land on earth that we had in 1960 (Wallace-Wells 2017).

[24] Brown 2009: 37ff.

[25] Diaz and Rosenberg 2008.

[26] As the Millennium Ecosystem Assessment (2005) reports: "approximately 60% (15 out of 24) of the ecosystem services examined ... are being degraded or used unsustainably, including fresh water, capture fisheries, air and water purification, and the regulation of regional and local climate, natural hazards, and pests. The full costs of the loss and degradation of these ecosystem services ... are substantial and growing. Many ecosystem services have been degraded as a consequence of actions taken to increase the supply of other services, such as food. These trade-offs often shift the costs of degradation from one group of people to another or defer costs to future generations."

[27] United Nations Department of Economic and Social Affairs 2022.

[28] Thiele 2013: 104.

to almost double."[29] Yet, as we approach the outer bounds of photosynthetic efficiency, there is little space left for additional agricultural productivity gains, beyond those afforded by (the, again, unsustainable) Green Revolution.[30] In other words, agricultural yields today are nearly as good as they possibly can be.[31] We therefore cannot depend on new technological advances to increase output.

Adding to this, an often overlooked but nonetheless serious complication of climate change is its effect on the *quality* of agricultural production. There is increasing evidence that, after a certain point, higher concentrations of atmospheric carbon dioxide (CO_2) negatively affect not only plant *growth*, leading to lower yields, but also plant *development*, in the form of "nutrient collapse."[32] Regarding the first issue, as the FAO reports:

> A recent multi-model study using IPCC's highest scenario of warming found a mean effect on yields of four crop groups (coarse grains, oil seeds, wheat and rice, accounting for about 70 percent of global crop harvested area) of minus 17 percent globally by 2050 relative to a scenario with unchanging climate ... After 2050, the risk of more severe impacts increases.[33]

This is a striking reversal of the long-held view that higher concentrations of CO_2 simply and straightforwardly promote plant growth. (This false idea is often played up by US climate deniers, like former Chair of the House Committee on Science, Representative Lamar Smith, who argued in a 2017 op-ed that a "higher concentration of carbon dioxide in our atmosphere would aid photosynthesis, which ... correlates to a greater volume of food production and better quality food."[34])

Regarding plant development, researchers have found that, after a certain threshold—which we have already begun to pass—higher CO_2 concentrations depress the production of essential nutrients and minerals in key staple crops, including wheat, rice, barley, and potatoes. For instance, according to the IPCC, "wheat grown at 546–586 ppm CO_2 has 5.9–12.7% less protein, 3.7–6.5% less zinc, and 5.2–7.5% less iron."[35] This has vast implications for human health. Researchers estimate that, because of nutrient collapse, by 2050:

> 150 million people could be put at risk of protein deficiency, particularly in countries like India and Bangladesh ... [A] loss of zinc, which is particularly essential for

[29] Food and Agriculture Organization 2009a, 2009b. Other estimates have been made, some more optimistic (for instance, van Dijk, et al. [2021] predict that food consumption will increase by only 45–56 percent), others more pessimistic (e.g. Tilman, et al. [2011] anticipate a "+100–110% increase in global crop demand from 2005 to 2050"). Of course, some have disputed the claim that increased consumption is necessary, though not always in compelling or well-evidenced ways (see e.g. Holt-Giménez and Altieri 2013 or Tomlinson 2013). Perhaps a more convincing basis for challenging growth projections would be to highlight the potential of a widespread transition to plant-based diets, as the IPCC at times appears to suggest (see e.g. IPCC 2022a: ch. 5, pp. 792, 837). See also Friel 2019.

[30] As Brown (2009: 8) explains, "the rise in grainland productivity dropped from 2.1 percent a year from 1950 to 1990, to 1.3 percent from 1990 to 2008."

[31] At least without converting still more wilderness areas to agricultural land, which would of course further exacerbate the climate crisis by reducing carbon sinks. Consider the devastation inflicted, in the name of increased agricultural output, on some of the world's last remaining rainforests in Indonesia and Brazil. According to one disturbing study in a leading natural science journal, if we maintain our current "rate of population growth and resource consumption, in particular forest consumption," we have a less than 10 percent chance ("in the most optimistic scenario") of surviving "an irreversible collapse of our civilization" within the next four decades (Bologna and Aquino 2020).

[32] Evich 2017.

[33] Gitz, et al. 2016: xi, see also 9, 31, 53.

[34] Smith 2017.

[35] IPCC 2018a: ch. 5, p. 439.

> maternal and infant health, could put 138 million people at risk ... [Moreover,] more than 1 billion mothers and 354 million children live in countries where [plant-based] dietary iron is projected to drop significantly, which could exacerbate the already widespread public health problem of anemia.[36]

Thus meeting healthy baselines will require greater consumption—and yet, as just noted, the same problem driving nutrient collapse (high atmospheric concentrations of GHGs) is also diminishing our ability to produce (more) food. Together, then, with population growth, declining agricultural yields and nutrient collapse threaten new and difficult-to-overcome instances of extreme material scarcity, and thus risk political catastrophe.

[36] Evich 2017.

References

Ackerly, Brooke A. 2018. *Just Responsibility: A Human Rights Theory of Global Justice.* New York: Oxford University Press.

Adams, Matthew, and Ross Mittiga. 2021. "Material Scarcity and Scalar Justice." *Philosophical Studies* 178: 2237–56.

Adams, N. P. 2017. "Uncivil Disobedience: Political Commitment and Violence." *Res Publica* 24: 475–91.

Afshin, Ashkan, Patrick John Sur, Kairsten A. Fay, Leslie Cornaby, Giannina Ferrara, Joseph S. Salama, Erin C. Mullany (GBD 2017 Diet Collaborators). 2019. "Health Effects of Dietary Risks in 195 Countries, 1990–2017: A Systematic Analysis for the Global Burden of Disease Study 2017." *The Lancet* 393: 1958–72.

Agamben, Giorgio. 2005. *State of Exception.* Translated by Kevin Attell. Chicago, IL: University of Chicago Press.

Ahmed, Azam, and Kirk Semple. 2017. "Desperation Mounts in Caribbean Islands: 'All the Food Is Gone.'" New York Times (September 10).

Aldridge, Robert, and Virginia Stark. 1986. "Nuclear War, Citizen Intervention, and the Necessity Defense." *Santa Clara Law Review* 26: 299–354.

Allen, Evan. 2013. "Anticoal Activists Block Path of Freighter." Boston Globe (May 15).

Anderson, David G., Albert C. Goodyear, James Kennett, and Allen West. 2011. "Multiple Lines of Evidence for Possible Human Population Decline/Settlement Reorganization During the Early Younger Dryas." *Quaternary International* 242: 570–83.

Archer, David. 2016. *The Long Thaw: How Humans Are Changing the Next 100,000 Years of Earth's Climate.* Princeton, NJ: Princeton University Press.

Arendt, Hannah. 1970. "Civil Disobedience." The New Yorker (September 12). 70–105.

Aristotle. 1998. *Politics.* Translated by C. D. C. Reeve. Indianapolis, IN: Hackett Publishing Company.

Armstrong, Chris. 2016. "Fairness, Free-Riding and Rainforest Protection." *Political Theory* 44: 106–30.

Armstrong, Chris. 2022. "Book Review: *How to Blow up a Pipeline* by Andreas Malm." Ethics, Policy & Environment. https://www.tandfonline.com/toc/cepe21/0/0.

Armstrong McKay, David I., Arie Staal, Jesse F. Abrams, Ricarda Winkelmann, Boris Sakschewski, Sina Loriani, et al. 2022. "Exceeding 1.5°C Global Warming Could Trigger Multiple Climate Tipping Points." *Science* 377: 7950.

Arneson, Richard. 1999. "Egalitarianism and Responsibility." *Journal of Ethics* 3: 225–47.

Arneson, Richard. 2000. "Luck Egalitarianism and Prioritarianism." *Ethics* 110: 339–49.

Arsenault, Chris. 2014. "Only 60 Years of Farming Left if Soil Degradation Continues." Scientific American (December 5).

Asch, Rebecca G. 2015. "Climate Change and Decadal Shifts in the Phenology of Larval Fishes in the California Current Ecosystem." *Proceedings of the National Academy of Sciences* 112: E4065–74.

Associated Press. 2018. "Charges Dropped against Climate Activists Who Tried to Shut Down Pipelines." The Guardian (October 10). https://www.theguardian.com/environment/2018/oct/09/climate-change-activists-necessity-defense-charges-dropped (accessed October 2, 2023).

Aton, Adam. 2017. "Once Again, Climate Change Cited as Trigger for Conflict." Scientific American (June 9).

Barnett, Zach. 2020. "Why You Should Vote to Change the Outcome." *Philosophy & Public Affairs* 48: 422–46.

Barro, Robert J. 2009. "Rare Disasters, Asset Prices, and Welfare Costs." *American Economic Review* 99: 243–64.

Barry, Brian. 1970. *Sociologists, Economists, and Democracy*. Chicago, IL: University of Chicago Press.

Barry, Brian. 1978. "Circumstances of Justice and Future Generations." In *Obligations to Future Generations*, ed. R. I. Sikora and Brian Barry. Philadelphia, PA: Temple University Press. 204–48.

Barry, Brian. 1999. "Sustainability and Intergenerational Justice." In *Fairness and Futurity*, ed. Andrew Dobson. Abingdon and New York: Oxford University Press. 93–117.

Barry, John. 2012. "Climate Change, 'the Cancer Stage of Capitalism' and the Return of Limits to Growth: Towards a Political Economy of Sustainability." In *Climate Change and the Crisis of Capitalism*, ed. Mark Pelling, David Manuel-Navarrete, and Michael Redclift. London and New York: Routledge. 129–42.

Barry, John. 2021. "Green Republicanism and a 'Just Transition' from the Tyranny of Economic Growth." *Critical Review of International Social and Political Philosophy* 24: 725–42.

Barry, John, and Robyn Eckersley. 2005. *The State and the Global Ecological Crisis*. Cambridge, MA: MIT Press.

Barry, John, and Kimberly Smith. 2008. "Civic Republicanism and Green Politics." In *Building a Citizen Society: The Emerging Politics of Republican Democracy*, ed. Stuart White and Daniel Leighton. London: Lawrence & Wishart. 123–45.

Battisti, David. S., and Rosamond L. Naylor. 2009. "Historical Warnings of Future Food Insecurity with Unprecedented Seasonal Heat." *Science* 323: 240–4.

Battistoni, Alyssa. 2018. "States of Emergency." The Nation (June 21).

Battistoni, Alyssa. 2022. "Is Sabotage a Pipe Dream?" Verso Books [blog], April 21, https://www.versobooks.com/blogs/5324-is-sabotage-a-pipe-dream (accessed October 7, 2022).

Bauer, Steven M., and Peter J. Eckerstrom. 1986. "The State Made Me Do It: The Applicability of the Necessity Defense to Civil Disobedience." *Stanford Law Review* 39: 1173–200.

BBC News. 1998. "Ethnic Chinese Tell of Mass Rapes." *BBC News*, June 23.

BBC News. 2018a. "Extinction Rebellion Protests Block London Bridges." *BBC News*, November 17.

BBC News. 2018b. "France Fuel Unrest: 'Shame' on Violent Protesters, Says Macron." *BBC News*, November 25. https://www.bbc.com/news/world-europe-46331783 (accessed September 18, 2023).

BBC News. 2022. "Just Stop Oil Protesters at Kingsbury Depot." *BBC News*, September 19.

Beard, S. J., Lauren Holt, Asaf Tzachor, Luke Kemp, Shahar Avin, Phil Torres, et al. 2021. "Assessing Climate Change's Contribution to Global Catastrophic Risk." *Futures* 127: 102673.

Bedau, Hugo A. 1961. "On Civil Disobedience." *Journal of Philosophy* 58: 653–65.

Beeson, Mark. 2010. "The Coming of Environmental Authoritarianism." *Environmental Politics* 19: 276–94.

Bell, Daniel A. 2016. *The China Model: Political Meritocracy and the Limits of Democracy*. Princeton, NJ: Princeton University Press.

Bender, Bryan. 2013. "Chief of US Pacific Forces Calls Climate Biggest Worry." Boston Globe (March 9).
Bettis, Oliver D., Simon Dietz, and Nick G. Silver. 2017. "The Risk of Climate Ruin." *Climatic Change* 140: 109–18.
Bidder 70. 2012. "About the Film." https://www.bidder70film.com/about (accessed January 13, 2021).
Bidgood, Jess. 2014. "Charges Dropped against Climate Activists." New York Times (September 8).
Biello, David. 2009. "Farmed Out: How Will Climate Change Impact World Food Supplies?" Scientific American (September 30).
Biesemans, Bart, and Clement Rossignol. 2023. "Tractors Roll into Brussels in Farmer Protest over Plans to Limit Nitrogen Emissions." Reuters, March 3.
Boffey, Daniel. 2021. "Court Orders Royal Dutch Shell to Cut Carbon Emissions by 45% by 2030." The Guardian (May 26).
Bologna, Mauro, and Gerardo Aquino. 2020. "Deforestation and World Population Sustainability: A Quantitative Analysis." *Scientific Reports* 10: 7631.
Booth, Carol. 2012. "Bystanding and Climate Change." *Environmental Values* 21: 397–416.
Bostrom, Nick. 2013. "Existential Risk Prevention as Global Priority." *Global Policy* 4: 15–31.
Boyce, Daniel G., Derek P. Tittensor, Cristina Garilao, Stephanie Henson, Kristin Kaschner, Kathleen Kesner-Reyes, et al. 2022. "A Climate Risk Index for Marine Life." *Nature Climate Change* 12: 854–62.
Bradshaw, Corey J. A., Paul R. Ehrlich, Andrew Beattie, Gerardo Ceballos, Eileen Crist, Joan Diamond, et al. 2021. "Underestimating the Challenges of Avoiding a Ghastly Future." *Frontiers in Conservation Science* 1: 615419.
Brennan Center for Justice. 2018. "A Guide to Emergency Powers and Their Use." Brennan Center for Justice at New York University School of Law. https://www.brennancenter.org/our-work/research-reports/guide-emergency-powers-and-their-use (accessed October 2, 2023).
Brennan, Jason. 2011. *The Ethics of Voting*. Princeton, NJ: Princeton University Press.
Bressler, R. Daniel. 2021. "The Mortality Cost of Carbon." *Nature Communications* 12: 4467.
Broome, John. 2010 "The Most Important Thing About Climate Change." In *Public Policy: Why Ethics Matters*, ed. Jonathan Boston, Andrew Bradstock and David Eng. (ANU E Press). 101–16. https://library.oapen.org/bitstream/handle/20.500.12657/33647/459474.pdf?sequence=1&isAllowed=y.
Broome, John. 2012. *Climate Matters: Ethics in a Warming World*. New York: Norton.
Broome, John. 2016. "Do Not Ask for Morality." In *The Ethical Underpinnings of Climate Economics*, ed. Adrian Walsh, Säde Hormio, and Duncan Purves. Abingdon and New York: Routledge. 9–21.
Broome, John. 2020. "Why Not a Carbon Tax?" Keynote lecture at the internet conference on "The Pitfalls and Promises of Taxing Carbon", Pontificia Universidad Católica and Erasmus University, December 2020. https://www.youtube.com/watch?v=3SOD1Bhy9zU.
Broome, John. 2021a. "Self-Interest against Climate Change." YouTube. https://www.youtube.com/watch?v=SeVOL4kBlHI (accessed October 2, 2023).
Broome, John. 2021b. "How Much Harm Does Each of Us Do?" In *Philosophy and Climate Change*, ed. Mark Budolfson, Tristram McPherson, and David Plunkett. Oxford and New York: Oxford University Press. 281–91.

Broome, John, and Duncan Foley. 2016. "A World Climate Bank." In *Institutions for Future Generations*, ed. Iñigo González-Ricoy and Axel Gosseries. Oxford: Oxford University Press. 156–69.

Brown, Lester R. 2009. *Plan B 4.0: Mobilizing to Save Civilization*. New York: W. W. Norton & Company.

Brownlee, Kimberley. 2012. *Conscience and Conviction: The Case for Civil Disobedience*. Oxford: Oxford University Press.

Brownlee, Kimberley. 2017. "Civil Disobedience." *Stanford Encyclopedia of Philosophy* (Fall 2017 edition), ed. Edward Zalta. https://plato.stanford.edu/archives/fall2017/entries/civil-disobedience/ (accessed October 2, 2023).

Brundtland, Gro Harlem, Paul Ehrlich, Jose Goldemberg, James Hansen, Amory Lovins, Gene Likens, and James Lovelock. 2012. *Environment and Development Challenges: The Imperative to Act*. Tokyo: Asahi Glass Foundation.

Brysse, Keynyn, Naomi Oreskes, Jessica O'Reilly, and Michael Oppenheimer. 2013. "Climate Change Prediction: Erring on the Side of Least Drama?" *Global Environmental Change* 23: 327–37.

Buchanan, Allen. 2002. "Political Legitimacy and Democracy." *Ethics* 112: 689–719.

Buchanan, N. H. 2004. "Social Security, Generational Justice, and Long-Term Deficits." *Tax Law Review* 58: 275–326.

Budolfson, Mark. 2021. "Political Realism, Feasibility Wedges, and Opportunities for Collective Action on Climate Change." In *Philosophy and Climate Change*, ed. Mark Budolfson, Tristram McPherson, and David Plunkett. Oxford and New York: Oxford University Press. 323–45.

Buhaug, Halvard, Tor A. Benjaminsen, Elisabeth A. Gilmore, and Cullen S. Hendrix. 2023. "Climate-Driven Risks to Peace over the 21st Century." *Climate Risk Management* 39: 100471.

Bulos, Nabih. 2016. "War-Torn Syria's Food Shortage May Lead to UN-Sponsored Humanitarian Airdrops." Los Angeles Times (June 2).

Burke, Marshall, Solomon M. Hsiang, and Edward Miguel. 2015. "Climate and Conflict." *Annual Review of Economics* 7: 577–617.

Burke, Marshall B., Edward Miguel, Shanker Satyanath, John A. Dykema, and David B. Lobell. 2009. "Warming Increases the Risk of Civil War in Africa." *Proceedings of the National Academy of Sciences* 106: 20670–4.

Burnett, Richard, Hong Chen, Mieczysław Szyszkowicz, Neal Fann, Bryan Hubbell, C. Arden Pope III, et al. 2018. "Global Estimates of Mortality Associated with Long-Term Exposure to Outdoor Fine Particulate Matter." *Proceedings of the National Academy of Sciences* 115: 9592–7.

Butzer, Karl W., and Georgina H. Endfield. 2012. "Critical Perspectives on Historical Collapse." *Proceedings of the National Academy of Sciences* 109: 3628–31.

Cagle, Susie. 2019. "'Protesters as Terrorists': Growing Number of States Turn Anti-Pipeline Activism into a Crime." The Guardian (July 8).

Callaway, Ewen. 2017. "Oldest *Homo Sapiens* Fossil Claim Rewrites Our Species' History." Nature News (June 7).

Caney, Simon. 2005. "Cosmopolitan Justice, Responsibility, and Global Climate Change." *Leiden Journal of International Law* 18: 747–75.

Caney, Simon. 2009a. "Climate Change and the Future: Discounting for Time, Wealth, and Risk." *Journal of Social Philosophy* 40: 163–86.

Caney, Simon. 2009b. "Human Rights, Responsibilities and Climate Change." In *Global Basic Rights*, ed. Charles Beitz and Robert Goodin. Oxford: Oxford University Press. 227–47.

Caney, Simon. 2010. "Climate Change and the Duties of the Advantaged." *Critical Review of International Social and Political Philosophy* 13: 203–28.

Caney, Simon. 2012. "Just Emissions." *Philosophy & Public Affairs* 40: 255–300.

Caney, Simon. 2014a. "Climate Change, Intergenerational Equity and the Social Discount Rate." *Politics, Philosophy & Economics* 13: 320–42.

Caney, Simon. 2014b. "Two Kinds of Climate Justice: Avoiding Harm and Sharing Burdens." *Journal of Political Philosophy* 22: 125–49.

Caney, Simon. 2016. "Political Institutions for the Future: A Fivefold Package." In *Institutions for Future Generations*, ed. Iñigo González-Ricoy and Axel Gosseries. Oxford: Oxford University Press. 135–55.

Capisani, Simona. 2020. "Territorial Instability and the Right to a Livable Locality." *Environmental Ethics* 42: 189–207.

Carleton, T., S. M. Hsiang, and M. Burke. 2016. "Conflict in a Changing Climate." *European Physical Journal: Special Topics* 225: 489–511.

Carroll, Rory. 2008. "Haiti: Mud Cakes Become Staple Diet as Cost of Food Soars Beyond a Family's Reach." The Guardian (July 28).

Casey, Michael. 2015. "Climate Change Could Drain Global Seafood Supplies." *CBS News*, July 2.

Castellana, Daniele, Sven Baars, Fred W. Wubs, and Henk A. Dijkstra. 2019. "Transition Probabilities of Noise-Induced Transitions of the Atlantic Ocean Circulation." *Scientific Reports* 9: 20284.

Celikates, Robin. 2013. "La Désobéissance Civile: Entre Nonviolence Et Violence." *Rue Descartes* 77: 35–51.

Celikates, Robin. 2016. "Democratizing Civil Disobedience." *Philosophy & Social Criticism* 42: 982–94.

Christiano, Thomas. 1996. *The Rule of the Many*. Boulder, CO: Westview Press.

Christiano, Thomas. 2008. *The Constitution of Equality*. Oxford and New York: Oxford University Press.

Civil Liberties Defense Center. 2023. "Cases." https://cldc.org/cases/ (accessed January 4, 2020).

Clark, Michael A., Nina G. G. Domingo, Kimberly Colgan, Sumil K. Thakrar, David Tilman, John Lynch, et al. 2020. "Global Food System Emissions Could Preclude Achieving the 1.5° and 2°C Climate Change Targets." *Science* 370: 705–8.

Climate Defense Project. 2018. "Principles of Climate Civil Resistance." https://climatedefenseproject.org/wp-content/uploads/2021/05/Principles-of-Climate-Civil-Resistance.pdf (accessed January 4, 2020).

Climate Defense Project. 2020. "Climate Necessity Defense Case Guide." https://climatedefenseproject.org/wp-content/uploads/2021/05/CDP-Climate-Necessity-Defense-Case-Guide.pdf (accessed December 29, 2020).

Climate Defense Project. 2021. "Political Necessity Defense Jurisdiction Guide." https://climatedefenseproject.org/wp-content/uploads/2022/08/Jurisdiction-Chart.pdf (accessed October 7, 2022).

Climate Disobedience Center. 2017a. "Necessity Strategy." https://www.climatedisobedience.org/necessity_strategy (accessed October 7, 2022).

Climate Disobedience Center. 2017b. "The Climate Necessity Defense: A Legal Tool for Climate Activists." https://www.climatedisobedience.org/necessitydefense (accessed October 2, 2023).

Climate Emergency Declaration. 2022. "Climate Emergency Declarations in 2,275 Jurisdictions and Local Governments Cover 1 Billion Citizens." https://climateemergencydeclaration.org/climate-emergency-declarations-cover-15-million-citizens/ (accessed 15 September 2022).

Coates, Ben. 2023. "Why Dutch Farmers Turned Their Flag Upside Down." New York Times (April 3).

Cohan, John Alan. 2007. "Civil Disobedience and the Necessity Defense." *Pierce Law Review* 6: 111–75.

Cohen, Chaim. 2011. "Eden." In *The Oxford Dictionary of the Jewish Religion*, eds. Adele Berlin and Maxine Grossman. New York: Oxford University Press. 228–9.

Cohen, Marshall. 1969. "Civil Disobedience in a Constitutional Democracy." *Massachusetts Review* 10: 211–26.

Cohen, Roger. 2022. "Of Barbecues and Men: A Summer Storm Brews over Virility in France." New York Times (September 5).

Cole, Matthew A. 2003. "Environmental Optimists, Environmental Pessimists and the Real State of the World—an Article Examining *The Skeptical Environmentalist: Measuring the Real State of the World* by Bjorn Lomborg." *Economic Journal* 113: F362–80.

Coogan, Michael David, Marc Zvi Brettler, Carol Ann Newsom, and Pheme Perkins. 2010. *The New Oxford Annotated Bible with Apocrypha: New Revised Standard Version*. Oxford and New York: Oxford University Press.

Cook, Benjamin I., Jason E. Smerdon, Richard Seager, and Sloan Coats. 2014. "Global Warming and 21st Century Drying." *Climate Dynamics* 23: 2607–27.

Cooper, J. H. 1986. "The Burden of the Public Debt: A Review." *South African Journal of Economics* 56: 278–91.

Copp, David. 1999. "The Idea of a Legitimate State." *Philosophy & Public Affairs* 28: 3–45.

Corbett, Jessica. 2019. "Washington Supreme Court Establishes 'Very Important Precedent' for Climate Necessity Defense in Case of Valve-Turner Ken Ward." Common Dreams, September 6. https://www.commondreams.org/news/2019/09/06/washington-supreme-court-establishes-very-important-precedent-climate-necessity (accessed October 2, 2023).

Cox, Daniel. 2017. "Way More Americans may be Atheists than we Thought." FiveThirtyEight, May 18. https://fivethirtyeight.com/features/way-more-americans-may-be-atheists-than-we-thought/ (accessed October 2, 2023).

Cui, Ying, Lee R. Kump, Andy J. Ridgwell, Adam J. Charles, Christopher K. Junium, Aaron F. Diefendorf, et al. 2011. "Slow Release of Fossil Carbon During the Palaeocene–Eocene Thermal Maximum." *Nature Geoscience* 4: 481–5.

Dahan, Maha El. 2016. "Syrian Food Crisis Deepens as War Chokes Farming." Reuters, April 26.

Dasgupta, Partha. 2008. "Discounting Climate Change." *Journal of Risk and Uncertainty* 37: 141–69.

Davenport, Coral. 2018. "Trump Administration Unveils Its Plan to Relax Car Pollution Rules." New York Times (August 2).

Davis, Mike. 2010. "Who Will Build the Ark?" *New Left Review* 61: 10–25.

Delmas, Candice. 2018. *A Duty to Resist: When Disobedience Should Be Uncivil*. New York: Oxford University Press.

Delmas, Candice, and Kimberley Brownlee. 2021. "Civil Disobedience." *Stanford Encyclopedia of Philosophy* (Summer 2021 edition), ed. Edward N. Zalta. https://plato.stanford.edu/archives/sum2021/entries/civil-disobedience/ (accessed October 2, 2023).

Delva, Joseph Guyler, and Jim Loney. 2008. "Haiti's Government Falls after Food Riots." Reuters, April 12.

Dennis, Brady. 2021. "Countries Must Ramp up Climate Pledges by 80 Percent to Hit Key Paris Target, Study Finds." Washington Post (February 21).

Derysh, Igor. 2019. "Trump's Repeal of Light Bulb Standards will Increase Pollution, Cost Billions." Salon, February 17. https://www.salon.com/2019/02/17/trumps-repeal-of-light-bulb-standards-will-increase-pollution-cost-billions/ (accessed October 2, 2023).

Di Paola, Marcello, and Dale Jamieson. 2018. "Climate Change and the Challenges to Democracy." *University of Miami Law Review* 72: 369–424.

Diamond, Jared. 2004. *Collapse: How Societies Choose to Fail or Succeed.* New York: Viking Press.

Diaz, Johnny. 2022. "Alaska Cancels Snow Crab Season Amid Population Declines." New York Times (October 14).

Diaz, Robert J., and Rutger Rosenberg. 2008. "Spreading Dead Zones and Consequences for Marine Ecosystems." *Science* 321: 926–9.

Ditlevsen, Peter, and Susanne Ditlevsen. 2023. "Warning of a Forthcoming Collapse of the Atlantic Meridional Overturning Circulation." *Nature Communications* 14: 4254.

Dobson, Andrew. 1996. "Representative Democracy and the Environment." In Democracy and the Environment, ed. W. M. Lafferty and J. Meadowcroft. Cheltenham: Edward Elgar. 124–39.

Duvat, Virginie K. E., Alexandre K. Magnan, Chris T. Perry, Tom Spencer, Johann D. Bell, Colette C. C. Wabnitz, et al. 2021. "Risks to Future Atoll Habitability from Climate-Driven Environmental Changes." *WIREs Climate Change* 12: e700.

Dworkin, Ronald. 1975. "The Original Position." In *Reading Rawls*, ed. Norman Daniels. New York: Basic Books. 16–52.

Eagleton, Terry. 2015. *Hope without Optimism.* New Haven, CT: Yale University Press.

Eckersley, Robyn. 2004. *The Green State: Rethinking Democracy and Sovereignty.* Cambridge, MA: MIT Press.

Eckersley, Robyn. 2015. "The Common but Differentiated Responsibilities of States to Assist and Receive 'Climate Refugees.'" *European Journal of Political Theory* 14: 481–500.

Eckersley, Robyn. 2023. "(Dis)Order and (in)Justice in a Heating World." *International Affairs* 99: 101–19.

Estlund, David. 2008. *Democratic Authority.* Princeton, NJ: Princeton University Press.

Estlund, David. 2016. "What is Circumstantial about Justice?" *Social Philosophy and Policy* 33: 292–311.

Evich, Helena Bottemiller. 2017. "The Great Nutrient Collapse." Politico, September 13. https://www.politico.com/agenda/story/2017/09/13/food-nutrients-carbon-dioxide-000511/ (accessed October 2, 2023).

Extinction Rebellion. 2022. "Declaration of Rebellion." https://extinctionrebellion.uk/declaration/ (accessed October 7, 2022).

Fassin, Didier, and Mariella Pandolfi. 2010. Contemporary States of Emergency: *The Politics of Military and Humanitarian Interventions.* New York: Zone Books.

Federman, Adam. 2020. "Revealed: US Listed Climate Activist Group as 'Extremists' Alongside Mass Killers." The Guardian (January 13).

Feldmann, Johannes, and Anders Levermann. 2015. "Collapse of the West Antarctic Ice Sheet after Local Destabilization of the Amundsen Basin." Proceedings of the National Academy of Sciences 112: 14191–6.

Ferguson, J. M. 1964. *Public Debt and Future Generations.* Chapel Hill, NC: University of North Carolina Press.

Fleurbaey, Marc, Maddalena Ferranna, Mark Budolfson, Francis Dennig, Kian Mintz-Woo, Robert Socolow, et al. 2018. "The Social Cost of Carbon: Valuing Inequality, Risk, and Population for Climate Policy." *The Monist* 102: 84–109.

Foley, D. 2007. *The Economic Fundamentals of Global Warming.* Santa Fe, NM: Santa Fe Instiute.

Food and Agriculture Organization. 2009a. "Global Agriculture Towards 2050." Paper presented at the High-level Expert Forum, Rome.

Food and Agriculture Organization. 2009b. *How to Feed the World in 2050.* Rome: FAO. https://www.fao.org/fileadmin/templates/wsfs/docs/Issues_papers/HLEF2050_Global_Agriculture.pdf.

Frame, Bob, and Nicholas A. Cradock-Henry. 2022. "Views from Nowhere, Somewhere and Everywhere Else: The Tragedy of the Horizon in the Early Anthropocene." *The Anthropocene Review* 10: 524–40.

Frederiksen, Kristian Vrede Skaaning. 2022. "Does Competence Make Citizens Tolerate Undemocratic Behavior?" *American Political Science Review* 116: 1–7.

Freedland, Jonathan. 2017. "In God We Trust: Why Americans Won't Vote in an Atheist President." The Guardian (August 8).

Friedrich, Johannes, Mengpin Ge, Andrew Pickens, and Leandro Vigna. 2020. "This Interactive Chart Shows Changes in the World's Top 10 Emitters." World Resources Institute. https://www.wri.org/insights/interactive-chart-shows-changes-worlds-top-10-emitters (accessed October 2, 2023).

Friel, Sharon. 2019. Climate Change and the People's Health, Small Books Big Ideas in Population Health. New York: Oxford University Press.

Furlong, Ashleigh. 2022. "Compromise Reached on Covid-19 Vaccine Intellectual Property Rights Waiver." Politico, March 15. https://www.politico.eu/article/compromise-reached-on-covid-19-vaccine-intellectual-property-rights-waiver/ (accessed October 2, 2023).

Furman, Jason, Rod Shadbegian, and Jim Stock. 2014. *The Cost of Delaying Action to Stem Climate Change.* Washington, DC: The White House.

Garcia-Gibson, Francisco. 2022. "Undemocratic Climate Protests." *Journal of Applied Philosophy* 39: 162–79.

García-Portela, Laura. 2020. "Moral Responsibility for Climate Change Loss and Damage." *Teorema: Revista Internacional de Filosofía* 39: 7–24.

Gardiner, Stephen M. 2004. "Ethics and Global Climate Change." *Ethics* 114: 555–600.

Gardiner, Stephen M. 2006. "A Core Precautionary Principle." *Journal of Political Philosophy* 14: 33–60.

Gardiner, Stephen M. 2011. *A Perfect Moral Storm: The Ethical Tragedy of Climate Change.* New York: Oxford University Press.

Gaus, Gerald. 1997. "Reason, Justification, and Consensus: Why Democracy Can't Have it All." In *Deliberative Democracy*, ed. James Bohman and William Rehg. Cambridge, MA: MIT Press. 205–42.

Gayle, Damien. 2022a. "Just Stop Oil Activists Stage Protests at Essex and Midlands Terminals." The Guardian (August 23).

Gayle, Damien. 2022b. "Just Stop Oil's 'Spring Uprising' Protests Funded by US Philanthropists." The Guardian (April 29).

Gilley, Bruce. 2012. "Authoritarian Environmentalism and China's Response to Climate Change." *Environmental Politics* 21: 287–307.

Gitz, Vincent, Alexandre Meybeck, L. Lipper, C. De Young, and S. Braatz. 2016. *Climate Change and Food Security: Risks and Responses.* Food and Agriculture Organization of the United Nations (FAO) Report. Rome: FAO.

Gleick, Peter H. 1991. "Environment and Security: The Clear Conn." *Bulletin of the Atomic Scientists* 47: 16–21.

Godard, Olivier. 2008. "The Stern Review on the Economics of Climate Change: Contents, Insights and Assessment of the Critical Debate." *SAPI EN [online] 1. S. Surveys and Perspectives Integrating Environment and Society* [online] 1.

Goitein, Elizabeth. 2019. "The Alarming Scope of the President's Emergency Powers." The Atlantic. https://www.theatlantic.com/magazine/archive/2019/01/presidential-emergency-powers/576418/ (accessed October 2, 2023).

Goldenberg, Suzanne. 2014. "Climate Change 'Already Affecting Food Supply'—UN." The Guardian (March 30).

González-Ricoy, Iñigo, and Axel Gosseries, eds. 2016. *Institutions for Future Generations.* Oxford: Oxford University Press.

Goodman, Amy. 2016. "Historic Trial Lets Activists Who Blocked Oil Train Cite Climate Change Threat in Their Defense." AlterNet, January 12. https://www.democracynow.org/2016/1/11/historic_trial_lets_activists_who_blocked.

Gowdy, John. 2020. "Our Hunter-Gatherer Future: Climate Change, Agriculture and Uncivilization." *Futures* 115: 102488.

Grandoni, Dino. 2022. "In a First, US Appoints a Diplomat for Plants and Animals." The Washington Post (September 29).

Greger, Michael, and Gene Stone. 2016. *How Not to Die: Discover the Foods Scientifically Proven to Prevent and Reverse Disease.* London: Pan Macmillan.

Guerrero, Alexander A. 2010. "The Paradox of Voting and the Ethics of Political Representation." *Philosophy & Public Affairs* 38: 272–306.

Haldon, J., L. Mordechai, T. P. Newfield, A. F. Chase, A. Izdebski, P. Guzowski, et al. 2018. "History Meets Palaeoscience: Consilience and Collaboration in Studying Past Societal Responses to Environmental Change." *Proceedings of the National Academy of Sciences USA* 115: 3210–18.

Han, Heejin. 2015. "Authoritarian Environmentalism under Democracy: Korea's River Restoration Project." *Environmental Politics* 24: 810–29.

Han, Heejin. 2017. "Singapore, a Garden City: Authoritarian Environmentalism in a Developmental State." *Journal of Environment & Development* 26: 3–24.

Hand, Mark. 2018. "Judge Shoots Down Climate Necessity Defense at Last Minute, yet Acquits Pipeline Protesters." Think Progress, October 10.

Harwatt, Helen, William J. Ripple, Abhishek Chaudhary, Matthew G. Betts, and Matthew N. Hayek. 2020. "Scientists Call for Renewed Paris Pledges to Transform Agriculture." *The Lancet Planetary Health* 4: e9–e10.

Hayden, Jones. 2023. "Dutch Farmers, Climate Activists Stage Protests in the Hague." Politico, March 12. https://www.politico.eu/article/netherlands-protests-hague-farmers-nitrogen-emissions-climate-activists-fossil-fuel-subsidies/#:~:text=Thousands%20of%20farmers%20protested%20in,state%20support%20for%20fossil%20fuels (accessed October 2, 2023).

Heath, Joseph. 2021. *Philosophical Foundations of Climate Change Policy*. New York: Oxford University Press.

Heine, Bernd. 1985. "The Mountain People: Some Notes on the Ik of North-Eastern Uganda." *Africa: Journal of the International African Institute* 55: 3–16.

Herf, Jeffrey. 2019. "Emergency Powers Helped Hitler's Rise. Germany Has Avoided Them Ever Since." The Washington Post (February 19).

Herrero, Mario, Benjamin Henderson, Petr Havlík, Philip K. Thornton, Richard T. Conant, Pete Smith, et al. 2016. "Greenhouse Gas Mitigation Potentials in the Livestock Sector." *Nature Climate Change* 6: 452–61.

Higham, James, Elisabeth Ellis, and James Maclaurin. 2019. "Tourist Aviation Emissions: A Problem of Collective Action." *Journal of Travel Research* 58: 535–48.

Hill, Thomas E. 2001. "Hypothetical Consent in Kantian Constructivism." *Social Philosophy and Policy* 18: 300–29.

Hobbes, Thomas. 1994. *Leviathan: With Selected Variants from the Latin Edition of 1668*, ed. Edwin Curley. Indianapolis, IN: Hackett.

Holden, Barry. 2002. *Democracy and Global Warming*. London: Continuum.

Holm, Sune, and Daniel Steel. 2019. "New Directions for the Precautionary Principle: Introduction." *Ethics, Policy & Environment* 22: 1–2.

Holt-Giménez, Eric, and Miguel A. Altieri. 2013. "Agroecology, Food Sovereignty, and the New Green Revolution." *Agroecology and Sustainable Food Systems* 37: 90–102.

Honegger, Matthias, Wil Burns, and David R. Morrow. 2021. "Is Carbon Dioxide Removal 'Mitigation of Climate Change'?" *Review of European, Comparative & International Environmental Law* 30: 327–35.

Howard, Jeffery. 2021. "Coronavirus Misinformation, Social Media, and Freedom of Speech." In *Political Philosophy in a Pandemic: Routes to a More Just Future*, ed. Fay Niker and Aveek Bhattacharya. London: Bloomsbury, 167–76

Hsiang, Solomon M., Marshall Burke, and Edward Miguel. 2013. "Quantifying the Influence of Climate on Human Conflict." *Science* 341: 1235367.

Huggel, Christian, Laurens M. Bouwer, Sirkku Juhola, Reinhard Mechler, Veruska Muccione, Ben Orlove, et al. 2022. "The Existential Risk Space of Climate Change." *Climatic Change* 174: 8.

Hume, David. 1751. An Enquiry Concerning the Principles of Morals. https://www.gutenberg.org/files/4320/4320-h/4320-h.htm.

IPCC. 2007. *Climate Change 2007: The Physical Science Basis: Contribution of Working Group I to the Fourth Assessment Report of the Intergovernmental Panel on Climate Change*, ed. S. Solomon, D. Qin, M. Manning, Z. Chen, M. Marquis, K. B. Averyt, et al. Cambridge, UK and New York: Cambridge University Press.

IPCC. 2013. *Climate Change 2013: The Physical Science Basis: Contribution of Working Group I to the Fifth Assessment Report of the Intergovernmental Panel on Climate Change*, ed. T. F. Stocker, D. Qin, G.-K. Plattner, M. Tignor, S. K. Allen, J. Boschung, et al. Cambridge, UK and New York: Cambridge University Press.

IPCC. 2014a. *Climate Change 2014: Impacts, Adaptation, and Vulerability: Contribution of Working Group II to the Fifth Assessment Report of the Intergovernmental Panel on Climate Change*, ed. C. B. Field, V. R. Barros, D. J. Dokken, K. J. Mach, M. D. Mastrandrea, T. E. Bilir, et al. Cambridge, UK and New York: Cambridge University Press.

IPCC. 2014b. *Climate Change 2014: Mitigation of Climate Change: Contribution of Working Group III to the Fifth Assessment Report of the Intergovernmental Panel on Climate Change*, ed. O. Edenhofer, R. Pichs-Madruga, Y. Sokona, E. Farahani, S. Kadner, K. Seyboth, et al. Cambridge, UK and New York: Cambridge University Press.

IPCC. 2018. "Mitigation Pathways Compatible with 1.5°C in the Context of Sustainable Development." In *Global Warming of 1.5°C: IPCC Special Report on the Impacts of Global Warming of 1.5°C above Pre-Industrial Levels and Related Global Greenhouse Gas Emission Pathways, in the Context of Strengthening the Global Response to the Threat of Climate Change, Sustainable Development, and Efforts to Eradicate Poverty*. Cambridge, UK: Cambridge University Press. 93–174.

IPCC. 2018a. *Climate Change and Land: An IPCC Special Report on Climate Change, Desertification, Land Degradation, Sustainable Land Management, Food Security, and Greenhouse Gas Fluxes in Terrestrial Ecosystems*, ed. P. R. Shukla, J. Skea, E. Calvo Buendia, V. Masson-Delmotte, H.-O. Pörtner, D. C. Roberts, et al. Cambridge, UK and New York: Cambridge University Press.

IPCC. 2018b. "Summary for Policymakers." In *Global Warming of 1.5°C: An IPCC Special Report on the Impacts of Global Warming of 1.5°C Above Pre-Industrial Levels and Related Global Greenhouse Gas Emission Pathways, in the Context of Strengthening the Global Response to the Threat of Climate Change, Sustainable Development, and Efforts to Eradicate Poverty*, ed. V. Masson-Delmotte, P. Zhai, H.-O. Pörtner, D. Roberts, J. Skea, P.R. Shukla, et al. Cambridge, UK and New York: Cambridge University Press. 3–24.

IPCC. 2018c. "Summary for Policymakers, Technical Summary, Frequently Asked Questions, Glossary." In *Global Warming of 1.5°C: An IPCC Special Report on the Impacts of Global Warming of 1.5°C Above Pre-Industrial Levels and Related Global Greenhouse Gas Emission Pathways, in the Context of Strengthening the Global Response to the Threat of Climate Change, Sustainable Development, and Efforts to Eradicate Poverty*, ed. V. Masson-Delmotte, P. Zhai, H.-O. Pörtner, D. Roberts, J. Skea, P.R. Shukla, et al. Cambridge, UK and New York: Cambridge University Press.

IPCC. 2021a. *Climate Change 2021: The Physical Science Basis: Contribution of Working Group I to the Sixth Assessment Report of the Intergovernmental Panel on Climate Change*, ed. V. Masson-Delmotte, P. Zhai, A. Pirani, S. L. Connors, C. Péan, S. Berger, et al. Cambridge, UK and New York: Cambridge University Press.

IPCC. 2021b. "Summary for Policymakers." In *Climate Change 2021: The Physical Science Basis: Contribution of Working Group I to the Sixth Assessment Report of the Intergovernmental Panel on Climate Change*, ed. V. Masson-Delmotte, P. Zhai, A. Pirani, S. L. Connors, C. Péan, S. Berger, et al. Cambridge, UK and New York: Cambridge University Press. 3–31.

IPCC. 2022a. *Climate Change 2022: Impacts, Adaptation, and Vulerability: Contribution of Working Group II to the Sixth Assessment Report of the Intergovernmental Panel on Climate Change*, ed. H.-O. Pörtner, D. C. Roberts, M. Tignor, E. S. Poloczanska, K. Mintenbeck, A. Alegría, et al. Cambridge, UK and New York: Cambridge University Press.

IPCC. 2022b. "Summary for Policymakers." In *Climate Change 2022: Mitigation of Climate Change: Contribution of Working Group III to the Sixth Assessment Report of the Intergovernmental Panel on Climate Change*, ed. P. R. Shukla, J. Skea, R. Slade, A. Al Khourdajie, R. van Diemen, D. McCollum, et al. Cambridge, UK and New York: Cambridge University Press. 3–48.

Irfan, Umair. 2018. "The EPA Refuted its own Bizarre Justification for Rolling Back Fuel Efficiency Standards." Vox, August 16. https://www.vox.com/2018/4/2/17181476/epa-fuel-economy-standards-tesla#:~:text=The%20EPA%20refuted%20its%20own,proposal%20would%20be%20more%20dangerous (accessed October 2, 2023).

Jamieson, Dale. 1992. "Ethics, Public Policy, and Global Warming." *Science, Technology, and Human Values* 17: 139–53.

Jamieson, Dale. 2014. *Reason in a Dark Time: Why the Struggle against Climate Change Failed—and What it Means for Our Future.* New York: Oxford University Press.

Jegarajah, Sri. 2012. "Recalling 'Tortilla Riots', Mexico President Warns About Food Crisis." CNBC, September 11.

Johansson, Emma Li, Marianela Fader, Jonathan W. Seaquist, and Kimberly A. Nicholas. 2016. "Green and Blue Water Demand from Large-Scale Land Acquisitions in Africa." *Proceedings of the National Academy of Sciences* 113: 11471–6.

John, Stephen. 2019. "The Politics of Certainty: The Precautionary Principle, Inductive Risk and Procedural Fairness." *Ethics, Policy & Environment* 22: 21–33.

Johnston, Emily. 2017. "I Shut Down an Oil Pipeline—Because Climate Change is a Ticking Bomb." The Guardian (November 24).

Just Stop Oil. 2022a. "51 Just Stop Oil Supporters Jailed after Mass Civil Resistance in Court." https://juststopoil.org/2022/09/15/51-just-stop-oil-supporters-jailed-after-mass-civil-resistance-in-court/ (accessed October 2, 2023).

Just Stop Oil. 2022b. "Campaign Background." https://juststopoil.org/background/ (accessed October 7, 2022).

Just Stop Oil. 2022c. "Just Stop Oil Tunnellers End Their 13 Day Occupation as Police Chief Acknowledges Climate Change a 'Serious Concern.'" https://juststopoil.org/2022/09/04/breaking-just-stop-oil-tunnellers-end-their-13-day-occupation-as-police-chief-acknowledges-climate-change-a-serious-concern/ (accessed October 2, 2023).

Kagan, Shelly. 2011. "Do I Make a Difference?" *Philosophy & Public Affairs* 39: 105–41.

Kareiva, Peter, and Valerie Carranza. 2018. "Existential Risk Due to Ecosystem Collapse: Nature Strikes Back." *Futures* 102: 39–50.

Kateb, George. 2002. "The Adequacy of the Canon." *Political Theory* 30: 482–505.

Katz, Jonathan M. 2008. "Poor Haitians Resort to Eating Dirt." National Geographic News (January 30).

Katzenstein, Peter J., and Lucia A. Seybert, eds. 2018. *Protean Power: Exploring the Uncertain and Unexpected in World Politics.* Cambridge, UK: Cambridge University Press.

Kelleher, J. Paul. 2015. "Is There a Sacrifice-Free Solution to Climate Change?" *Ethics, Policy & Environment* 18: 68–78.

Kelley, Colin P., Shahrzad Mohtadi, Mark A. Cane, Richard Seager, and Yochanan Kushnir. 2015. "Climate Change in the Fertile Crescent and Implications of the Recent Syrian Drought." *Proceedings of the National Academy of Sciences* 112: 3241–6.

Kemp, Luke, Chi Xu, Joanna Depledge, Kristie L. Ebi, Goodwin Gibbins, Timothy A. Kohler, et al. 2022. "Climate Endgame: Exploring Catastrophic Climate Change Scenarios." *Proceedings of the National Academy of Sciences* 119: e2108146119.

Kerry, John, and Ro Khanna. 2019. "Don't Let China Win the Green Race." New York Times (December 9).

Khazendar, Ala, Eric Rignot, Dustin M. Schroeder, Helene Seroussi, Michael P. Schodlok, Bernd Scheuchl, et al. 2016. "Rapid Submarine Ice Melting in the Grounding Zones of Ice Shelves in West Antarctica." *Nature Communications* 7: 13243.

King, Martin Luther. 1992. "Letter from Birmingham Jail." *UC Davis Law Review* 26: 835–51.

Kingston, Ewan. 2014. "Climate Justice and Temporally Remote Emissions." *Social Theory and Practice* 40: 281–303.

Klein, Naomi. 2014. *This Changes Everything: Capitalism vs. The Climate.* New York: Simon & Schuster.

Klosko, George. 1990. "Parfit's Moral Arithmetic and the Obligation to Obey the Law." *Canadian Journal of Philosophy* 20: 191–213.

Klosko, George. 2013. *History of Political Theory: An Introduction*. Vol. II: *Modern*. Oxford: Oxford University Press.

Kornhuber, Kai, Corey Lesk, Carl F. Schleussner, Jonas Jägermeyr, Peter Pfleiderer, and Radley M. Horton. 2023. "Risks of Synchronized Low Yields are Underestimated in Climate and Crop Model Projections." *Nature Communications* 14: 3528.

Korsgaard, Christine. 1997. *The Sources of Normativity*. Cambridge, UK: Cambridge University Press.

Korte, Gregory. 2017. "A Permanent Emergency: Trump Becomes Third President to Renew Extraordinary Post-9/11 Powers." USA Today, (September 14).

Kreuder-Sonnen, Christian. 2019. *Emergency Powers of International Organizations: Between Normalization and Containment*. Oxford and New York: Oxford University Press.

Kuran, Timur. 1991. "Now Out of Never: The Element of Surprise in the East European Revolution of 1989." *World Politics* 44: 7–48.

Kusnetz, Nicholas. 2019. "More States Crack Down on Pipeline Protesters, Including Supporters Who aren't Even on the Scene." Inside Climate News, March 28. https://insideclimatenews.org/news/28032019/pipeline-protest-crackdown-state-law-legislation-south-dakota-conspirators-riot-boosting/ (accessed October 2, 2023).

Labonte, M., and G. E. Makinen. 2005. *The National Debt: Who Bears its Burden?* Washington, DC: Congressional Research Service.

Lai, Ten-Herng. 2020. "Civil Disobedience, Costly Signals, and Leveraging Injustice." *Ergo: An Open Access Journal of Philosophy* 7: 1083–108.

Lai, Ten-Herng, and Chong-Ming Lim. (forthcoming). "Environmental Activism and the Fairness of Costs Argument for Uncivil Disobedience." *Journal of the American Philosophical Association* 9: 490–509.

Lanchester, John. 2017. "The Case against Civilization." *John Lanchester*, (September 28). https://www.newyorker.com/magazine/2017/09/18/the-case-against-civilization.

Lane, Melissa. 2018. "Uncertainty, Action and Politics: The Problem of Negligibility." In *Nature, Action and the Future: Political Thought and the Environment*, ed. Katrina Forrester and Sophie Smith. Cambridge, UK: Cambridge University Press. 157–79.

Laville, Sandra. 2020. "Ban SUV Adverts to Meet UK Climate Goals, Report Urges." The Guardian (August 3).

Lawrence, Frank. 1988. "The Nuremberg Principles: A Defense for Political Protesters." *Hastings Law Journal* 40: 397–436.

Lazar, Nomi Claire. 2009. *States of Emergency in Liberal Democracies*. New York: Cambridge University Press.

Lear, Jonathan. 2006. *Radical Hope: Ethics in the Face of Cultural Devastation*. Cambridge, MA: Harvard University Press.

Lefebvre, Ben, and Anthony Adragna. 2019. "Trump Administration Seeks Criminal Crackdown on Pipeline Protests." Politico, June 3. https://www.politico.com/story/2019/06/03/trump-administration-seeks-criminal-crackdown-on-pipeline-protests-1499008 (accessed October 2, 2023).

Lefkowitz, David. 2007. "On a Moral Right to Civil Disobedience." *Ethics* 117: 202–33.

Lemmons, John, and Donald A. Brown. 2011. "Global Climate Change and Non-Violent Civil Disobedience." *Ethics in Science and Environmental Politics* 11: 3–12.

Lenton, Timothy M., Johan Rockström, Owen Gaffney, Stefan Rahmstorf, Katherine Richardson, Will Steffen, et al. 2019. "Climate Tipping Points—Too Risky to Bet Against." *Nature* 575: 592–5.

Leopold, Aldo. 1989. *A Sand County Almanac, and Sketches Here and There*. New York: Oxford University Press.

Levin, Sam. 2017. "Judge in Environmental Activist's Trial Says Climate Change is Matter of Debate." The Guardian (January 31).

Lichtenberg, Judith. 2010. "Negative Duties, Positive Duties, and the 'New Harms.'" *Ethics* 120: 557–78.

Lind, R. C. 1995. "Intergenerational Equity, Discounting, and the Role of Cost–Benefit Analysis in Evaluating Global Climate Policy." *Energy Policy* 23: 379–89.

Liu, Peiran R., and Adrian E. Raftery. 2021. "Country-Based Rate of Emissions Reductions Should Increase by 80% Beyond Nationally Determined Contributions to Meet the 2°C Target." *Communications Earth & Environment* 2: 29.

Livingston, Alexander. 2018. "Fidelity to Truth: Gandhi and the Genealogy of Civil Disobedience." *Political Theory* 46: 511–36.

Lobster Boat Blockade. 2013. "The Action." https://lobsterboatblockade.org/the-action/ (accessed September 9, 2021).

Lobster Boat Blockade. 2014. "Lobster Boat Blockade 3: Live at Brayton Point on 5/15/13." YouTube. https://www.youtube.com/watch?v=X525OyACcCo (accessed October 2, 2023).

Lomborg, Bjørn. 2001. *The Skeptical Environmentalist: Measuring the Real State of the World*. Cambridge, UK: Cambridge University Press.

Lomborg, Bjørn. 2007. *Cool It: The Skeptical Environmentalist's Guide to Global Warming*. New York: Knopf.

Long, Lance N., and Ted Hamilton. 2018. "The Climate Necessity Defense: Proof and Judicial Error in Climate Protest Cases." *Stanford Environmental Law Journal* 38: 57–115.

Lu, Denise. 2019. "Trump's Emergency Declaration." New York Times (February 15).

Lüthi, D., M. Le Floch, B. Bereiter, T. Blunier, J.-M. Barnola, U. Siegenthaler, et al. 2008. "High-Resolution Carbon Dioxide Concentration Record 650,000-800,000 Years before Present." *Nature* 453: 379–82.

Machiavelli, Niccolò. 1994. *Machiavelli: Selected Political Writings*. Translated by David Wootton, ed. David Wootton. Indianapolis, IN: Hackett.

Maffly, Brian. 2018. "10 Years after He Monkey-Wrenched a Utah Oil and Gas Lease Auction, Tim DeChristopher is 'Feeling Demoralized' by 'the State of the World' but Sees Hope in Humanity." The Salt Lake Tribune (December 15).

Malm, Andreas. 2016. *Fossil Capital: The Rise of Steam Power and the Roots of Global Warming*. New York: Verso.

Malm, Andreas. 2021. *How to Blow up a Pipeline*. New York: Verso Books.

Mann, Geoff, and Joel Wainwright. 2018. *Climate Leviathan: A Political Theory of Our Planetary Future*. London: Verso.

Manson, Neil A. 2002. "Formulating the Precautionary Principle." *Environmental Ethics* 24: 263–74.

Maring, Luke. 2016. "Debate: Why Does the Excellent Citizen Vote?" *Journal of Political Philosophy* 24: 245–57.

Marques, António, Maria Leonor Nunes, Stephanie K. Moore, and Mark S. Strom. 2010. "Climate Change and Seafood Safety: Human Health Implications." *Food Research International* 43: 1766–79.

Martin, Adrienne. 2013. *How We Hope*. Princeton, NJ: Princeton University Press.

Martin, Ian W. R., and Robert S. Pindyck. 2021. "Welfare Costs of Catastrophes: Lost Consumption and Lost Lives." *Economic Journal* 131: 946–69.

Mazza, Patrick. 2016. "Why I Blocked an Oil Train—and Why You Should Engage in Climate Disobedience Too." Grist, January 27. https://grist.org/article/why-i-blocked-an-oil-train-and-why-you-should-engage-in-climate-disobedience-too/ (accessed October 2, 2023).

McGeown, Calum. 2023. "Revolutionary Green Republicanism: Anti-Capitalist Politics for the Planetary Crisis." Doctoral thesis, School of History, Anthropology, Philosophy and Politics, Queen's University Belfast.

McInerney, Francesca A., and Scott L. Wing. 2011. "The Paleocene–Eocene Thermal Maximum: A Perturbation of Carbon Cycle, Climate, and Biosphere with Implications for the Future." *Annual Review of Earth and Planetary Sciences* 39: 489–516.

McKibben, Bill. 2016. "A World at War." The New Republic (August 15). 22–31.

McKinnon, Catriona. 2012. *Climate Change and Future Justice: Precaution, Compensation, and Triage.* Abingdon and New York: Routledge.

McKinnon, Catriona. 2017. "Endangering Humanity: An International Crime?" *Canadian Journal of Philosophy* 47: 395–415.

Meckled-Garcia, Saladin. 2014. "What Comes First, Democracy or Human Rights?" *Critical Review of International Social and Political Philosophy* 17: 681–8.

Mehrabi, Zia. 2020. "Food System Collapse." *Nature Climate Change* 10: 16–17.

Meyer, Lukas H., and Dominic Roser. 2009. "Enough for the Future." In *Intergenerational Justice*, ed. Axel Gosseries and Lukas H. Meyer. Oxford: Oxford University Press. 219–48.

Middleton, Guy D. 2017. *Understanding Collapse: Ancient History and Modern Myths.* Cambridge, UK: Cambridge University Press.

Mill, John Stuart. 1998. *On Liberty and Other Essays.* New York: Oxford University Press.

Millennium Ecosystem Assessment. 2005. *Ecosystems and Human Well-Being: Synthesis.* Washington, DC: Island Press.

Mingle, Jonathan. 2008. "The Climate-Change Defense." New York Times Magazine, (December 12).

Minogue, Kristen. 2014. "Climate Change Expected to Expand Majority of Ocean Dead Zones." Smithsonian Insider (November 10).

Mittiga, Ross. unpublished. "Two Ways of Parsing Disobedience."

Mittiga, Ross. 2019a. "Allocating the Burdens of Climate Action: Consumption-Based Carbon Accounting and the Polluter-Pays Principle." In *Transformative Climates and Accountable Governance*, ed. Beth Edmondson and Stuart Levy, Palgrave Studies in Environmental Transformation, Transition and Accountability. Cham, Switzerland: Palgrave Macmillan. 157–94.

Mittiga, Ross. 2019b. "What's in a World? Du Bois and Heidegger on Politics, Aesthetics, and Foundings." *Contemporary Political Theory* 18: 180–201.

Mittiga, Ross. 2019c. "What's the Problem with Geo-Engineering?" *Social Theory and Practice* 45: 471–99.

Mittiga, Ross. 2022. "Political Legitimacy, Authoritarianism, and Climate Change." *American Political Science Review* 116: 998–1011.

Moellendorf, Darrel. 2022. *Mobilizing Hope: Climate Change and Global Poverty.* New York: Oxford University Press.

Morello, Lauren. 2012. "How Coral Bleaching Could Lead to Famine." Scientific American (April 3).

Morton, Adam, and Lisa Cox. 2022. "'Devastating': 91% of Reefs Surveyed on Great Barrier Reef Affected by Coral Bleaching in 2022." The Guardian (May 10).

Moskop, John C., and Kenneth V. Iserson. 2007. "Triage in Medicine, Part II: Underlying Values and Principles." *Annals of Emergency Medicine* 49: 282–7.

Mulgan, Tim. 2014. *Ethics for a Broken World: Imagining Philosophy after Catastrophe.* Abingdon and New York: Routledge.

Munthe, Christian. 2019. "The Black Hole Challenge: Precaution, Existential Risks and the Problem of Knowledge Gaps." *Ethics, Policy & Environment* 22: 49–60.

Murtaugh, Paul A., and Michael G. Schlax. 2009. "Reproduction and the Carbon Legacies of Individuals." *Global Environmental Change* 19: 14–20.

Nagel, Thomas. 1987. "Moral Conflict and Political Legitimacy." *Philosophy and Public Affairs* 16: 215–40.

National Research Council. 2011. *Climate Stabilization Targets: Emissions, Concentrations, and Impacts over Decades to Millennia.* Washington, DC: National Academies Press.

National Research Council. 2013. *Abrupt Impacts of Climate Change: Anticipating Surprises.* Washington, DC: National Academies Press.

Nijhuis, Michelle. 2018. "'I'm Just More Afraid of Climate Change Than I Am of Prison.'" New York Times Magazine (February 13).

Nordhaus, William D. 1997. "Discounting in Economics and Climate Change." *Climatic Change* 37: 315–28.

Nordhaus, William D. 2008a. *A Question of Balance: Weighing the Options on Global Warming Policies.* New Haven, CT: Yale University Press.

Nordhaus, William D. 2008b. "The Question of Global Warming: An Exchange." New York Review of Books (September 25).

Norgaard, Kari Marie. 2011. *Living in Denial: Climate Change, Emotions, and Everyday Life.* Cambridge, MA: MIT Press.

Nozick, Robert. 1974. *Anarchy, State, and Utopia.* New York: Basic Books.

NPR Staff. 2019. "Transcript: Greta Thunberg's Speech at the UN Climate Action Summit." NPR, September 23. https://www.npr.org/2019/09/23/763452863/transcript-greta-thunbergs-speech-at-the-u-n-climate-action-summit (accessed October 2, 2023).

Nuccitelli, Dana. 2017. "Nato Joins the Pentagon in Deeming Climate Change a Threat Multiplier." Bulletin of the Atomic Scientists, May 25. https://thebulletin.org/2017/05/nato-joins-the-pentagon-in-deeming-climate-change-a-threat-multiplier/ (accessed October 2, 2023).

Nussbaum, Martha. 2000. "Aristotle, Politics, and Human Capabilities. A Response to Anony, Arneson, Charlesworthy, and Mulgan." *Ethics* 111: 102–40.

Ophuls, William. 1973. "Leviathan, or Oblivion?" In *Toward a Steady State Economy*, ed. H. Daly. San Francisco: W. H. Freeman. 215–30.

Oreskes, Naomi, and Erik M. Conway. 2011. *Merchants of Doubt: How a Handful of Scientists Obscured the Truth on Issues from Tobacco Smoke to Global Warming.* New York: Bloomsbury USA.

Oreskes, Naomi, and Erik M. Conway. 2014. *The Collapse of Western Civilization: A View from the Future.* New York: Columbia University Press.

Oreskes, Naomi, Michael Oppenheimer, and Dale Jamieson. 2019. "Scientists Have Been Underestimating the Pace of Climate Change." Scientific American (August 19).

Page, Edward A. 2016. "Qui Bono? Justice in the Distribution of the Benefits and Burdens of Avoided Deforestation." *Res Publica* 22: 83–97.

Painting, Rob. 2011. "Co2 Currently Rising Faster Than the PETM Extinction Event." Post on Skeptical Science, June 17. https://skepticalscience.com/co2-rising-ten-times-faster-than-petm-extinction.html (accessed October 2, 2023).

Panggabean, Samsu Rizal, and Benjamin Smith. 2011. "Explaining Anti-Chinese Riots in Late 20th Century Indonesia." *World Development* 39: 231–42.

Parfit, Derek. 1984. *Reasons and Persons*. Oxford: Oxford University Press.

Pashley, Alex. 2015. "Climate Change Migration is 'Genocide', Says Marshall Islands Minister." Climate Home News (May 10).

Pasternak, Avia. 2018. "Political Rioting: A Moral Assessment." *Philosophy & Public Affairs* 46: 384–418.

Pateman, Carole. 1988. *The Sexual Contract*. Stanford, CA: Stanford University Press.

Peel, Michael, and Javier Blas. 2013. "Syria's Food Shortages Worsening, UN Says." The Washington Post (January 23).

Penman, Donald E., Barbel Hönisch, Richard E. Zeebe, Ellen Thomas, and James C. Zachos. 2014. "Rapid and Sustained Surface Ocean Acidification During the Paleocene–Eocene Thermal Maximum." *Paleoceanography* 29: 357–69.

Peter, Fabienne. 2017. "Political Legitimacy." *Stanford Encyclopedia of Philosophy* (Summer 2017 edition), ed. Edward N. Zalta. https://plato.stanford.edu/archives/sum2017/entries/legitimacy/ (accessed October 2, 2023).

Pineda, Erin. 2015. "Civil Disobedience and Punishment: (Mis)Reading Justification and Strategy from SNCC to Snowden." *History of the Present* 5: 1–30.

Plumer, Brad, and Nadja Popovich. 2021. "Yes, There Has Been Progress on Climate. No, It's Not Nearly Enough." New York Times (October 21).

Pölzler, Thomas. 2021. "Basic Needs in Normative Contexts." *Philosophy Compass* 16: e12732.

Popper, Karl. 1945. *The Open Society and Its Enemies*. London: Routledge.

Posner, Eric A., and David Weisbach. 2010. *Climate Change Justice*. Princeton, NJ: Princeton University Press.

Posner, Richard A. 2004. *Catastrophe: Risk and Response*: Oxford University Press.

Povitkina, Marina. 2018. "The Limits of Democracy in Tackling Climate Change." *Environmental Politics* 27: 411–32.

Purdey, Jemma. 2006. *Anti-Chinese Violence in Indonesia, 1996–1999*. Honolulu: University of Hawaii Press.

Queally, Jon. 2013. "The Lobster Boat vs. the Coal Freighter as Climate Activists Blockade Power Plant." Common Dreams, May 15. https://www.commondreams.org/news/2013/05/15/lobster-boat-vs-coal-freighter-climate-activists-blockade-power-plant (accessed October 2, 2023).

Rahmstorf, Stefan, Jason E. Box, Georg Feulner, Michael E. Mann, Alexander Robinson, Scott Rutherford, et al. 2015. "Exceptional Twentieth-Century Slowdown in Atlantic Ocean Overturning Circulation." *Nature Climate Change* 5: 475–80.

Rausch, Joseph. 2019. "The Necessity Defense and Climate Change: A Climate Change Litigant's Guide." *Columbia Journal of Environmental Law* 44: 553–602.

Rawls, John. 1999. *A Theory of Justice*, revised edition. Cambridge, MA: Harvard University Press.

Rawls, John. 2005. *Political Liberalism*, expanded edition. New York: Columbia University Press.

Ray, Deepak K., Nathaniel D. Mueller, Paul C. West, and Jonathan A. Foley. 2013. "Yield Trends are Insufficient to Double Global Crop Production by 2050." *PLOS ONE* 8: e66428.

Raz, Joseph. 1995. *The Morality of Freedom*. Oxford and New York: Oxford University Press.

Raz, Joseph. 2009. *The Authority of Law: Essays on Law and Morality*. New York: Oxford University Press.
Rendall, Matthew. 2011. "Climate Change and the Threat of Disaster: The Moral Case for Taking out Insurance at Our Grandchildren's Expense." *Political Studies* 59: 884–99.
Reuters. 2017. "Syrian War Monitor Says 465,000 Killed in Six Years of Fighting." Reuters, March 13.
Richards, C. E., H. L. Gauch, and J. M. Allwood. 2023. "International Risk of Food Insecurity and Mass Mortality in a Runaway Global Warming Scenario." *Futures* 150: 103173.
Richards, C. E., R. C. Lupton, and J. M. Allwood. 2021. "Re-Framing the Threat of Global Warming: An Empirical Causal Loop Diagram of Climate Change, Food Insecurity and Societal Collapse." *Climatic Change* 164: 49.
Riofrancos, Thea. 2022. "A Burning Planet: Should the Climate Movement Embrace Sabotage?" The Nation (July 25).
Ripple, William J., Pete Smith, Helmut Haberl, Stephen A. Montzka, Clive McAlpine, and Douglas H. Boucher. 2014. "Ruminants, Climate Change and Climate Policy." *Nature Climate Change* 4: 2–5.
Ripple, William J., Christopher Wolf, Thomas M. Newsome, Phoebe Barnard, and William R. Moomaw. 2022. "World Scientists' Warning of a Climate Emergency." *BioScience* 72: 1149–55.
Ripstein, Arthur. 1998. *Equality, Responsibility, and the Law*. Cambridge, UK: Cambridge University Press.
Ripstein, Arthur. 2004. "Authority and Coercion." *Philosophy and Public Affairs* 32: 2–35.
Ritchie, Hannah, Max Roser, and Pablo Rosado. 2020. "CO_2 and Greenhouse Gas Emissions." Our World in Data. https://ourworldindata.org/co2-and-greenhouse-gas-emissions (accessed October 2, 2023).
Roberts, David. 2018. "Washington Votes No on a Carbon Tax—Again." Vox, November 6. https://www.vox.com/energy-and-environment/2018/9/28/17899804/washington-1631-results-carbon-fee-green-new-deal (accessed October 2, 2023).
Robeyns, Ingrid. 2017. "Having Too Much." *NOMOS: American Society for Political and Legal Philosophy* 58: 1–44.
Robinson, Kim Stanley. 2020. *The Ministry for the Future: A Novel*. New York: Orbit Books.
Rockström, Johan, Will Steffen, Kevin Noone, Åsa Persson, F. Stuart Chapin, Eric F. Lambin, et al. 2009. "A Safe Operating Space for Humanity." *Nature* 461: 472–5.
Romanello, Marina, Alice McGushin, Claudia Di Napoli, Paul Drummond, Nick Hughes, Louis Jamart, et al. 2021. "The 2021 Report of the Lancet Countdown on Health and Climate Change: Code Red for a Healthy Future." *The Lancet* 398: 1619–62.
Roser, Dominic. 2020. "The Case for Climate Hope." In *Politik Der Zukunft*, ed. Tamoudi Nejma, Faets Simon, and Reder Michael. Bielefeld: Transcript Verlag. 65–86.
Rossiter, Clinton. 1948. *Constitutional Dictatorship: Crisis Government in the Modern Democracies*. Princeton, NJ: Princeton University Press.
Rulli, Maria Cristina, Antonio Savioria, and Paolo D'Odorico. 2013. "Global Land and Water Grabbing." *Proceedings of the National Academy of Sciences* 110: 892–7.
Sandin, Per. 1999. "Dimensions of the Precautionary Principle." *Human & Ecological Risk Assessment* 5: 889–907.
Sandin, Per, and Martin Peterson. 2019. "Is the Precautionary Principle a Midlevel Principle?" *Ethics, Policy & Environment* 22: 34–48.
Scarborough, Peter, Michael Clark, Linda Cobiac, Keren Papier, Anika Knuppel, John Lynch, et al. 2023. "Vegans, Vegetarians, Fish-Eaters and Meat-Eaters in the UK Show Discrepant Environmental Impacts." *Nature Food* 4: 565–74.

Scharpf, Fritz. 1999. *Governing in Europe: Effective and Democratic?*. Oxford and New York: Oxford University Press.

Scheffler, Samuel. 2016. *Death and the Afterlife*. New York: Oxford University Press.

Schelling, Thomas C. 2000. "Intergenerational and International Discounting." *Risk Analysis* 20: 833–8.

Scheuerman, William E. 2006. "Survey Article: Emergency Powers and the Rule of Law after 9/11." *Journal of Political Philosophy* 14: 61–84.

Scheuerman, William E. 2014. "Whistleblowing as Civil Disobedience: The Case of Edward Snowden." *Philosophy & Social Criticism* 40: 609–28.

Scheuerman, William E. 2015. "Recent Theories of Civil Disobedience: An Anti-Legal Turn?" *Journal of Political Philosophy* 23: 427–49.

Scheuerman, William E. 2019. "Why Not Uncivil Disobedience?" *Critical Review of International Social and Political Philosophy* 25: 980–99.

Scheuerman, William E. 2022. "Political Disobedience and the Climate Emergency." *Philosophy & Social Criticism* 48: 791–812.

Schlenker, Wolfram, and Michael J. Roberts. 2009. "Nonlinear Temperature Effects Indicate Severe Damages to US Crop Yields under Climate Change." *Proceedings of the National Academy of Sciences* 106: 15594–8.

Schmitt, Carl. 1985. *Political Theology: Four Chapters on the Concept of Sovereignty*. Translated by George Schwab. Cambridge, MA: MIT Press.

Schulman, Jeremy. 2013. "Defense Contractor: Climate Change Could Create 'Business Opportunities.'" Mother Jones, August 14. https://www.motherjones.com/environment/2013/08/raytheon-climate-change-security/#:~:text=The%20Massachusetts%2Dbased%20defense%20contractor,gunning%20for%20catastrophic%20global%20warming (accessed October 2, 2023).

Scott, James C. 2017. *Against the Grain: A Deep History of the Earliest States*. New Haven, CT: Yale University Press.

Scranton, Roy. 2015. *Learning to Die in the Anthropocene: Reflections on the End of a Civilization*. San Francisco, CA: City Lights Publishers.

Seaquist, J. W., Emma Li Johansson, and Kimberly A. Nicholas. 2014. "Architecture of the Global Land Acquisition System: Applying the Tools of Network Science to Identify Key Vulnerabilities." *Environmental Research Letters* 9: 114006.

Sen, Amartya. 1984. *Resources, Values, and Development*. Cambridge, MA: Harvard University Press.

Shiffrin, S. 1999. "Wrongful Life, Procreative Responsibility, and the Significance of Harm." *Legal Theory* 5: 117–48.

Shue, Henry. 1996. *Basic Rights: Subsistence, Affluence, and US Foreign Policy*. Princeton, NJ: Princeton University Press.

Shue, Henry. 1999. "Global Environment and International Inequality." *International Affairs* 75: 531–45.

Shue, Henry. 2010. "Deadly Delays, Saving Opportunities: Creating a More Dangerous World?" In *Climate Ethics: Essential Readings*, ed. Stephen M. Gardiner, Simon Caney, Dale Jamieson, and Henry Shue. New York: Oxford University Press. 146–62.

Simmons, A. John. 1999. "Justification and Legitimacy." *Ethics* 109: 739–71.

Simmons, A. John. 2001. *Justification and Legitimacy: Essays on Rights and Obligations*. Cambridge, UK: Cambridge University Press.

Singer, Peter. 1972. "Famine, Affluence, and Morality." *Philosophy & Public Affairs* 1: 229–43.

Singer, Peter. 2010. "One Atmosphere." In *Climate Ethics: Essential Readings*, ed. Stephen M. Gardiner, Simon Caney, Dale Jamieson, and Henry Shue. New York: Oxford University Press. 181–99.

Smith, Rep. Lamar. 2017. "Don't Believe the Hysteria over Carbon Dioxide." The Daily Signal, July 24. https://www.dailysignal.com/2017/07/24/dont-believe-hysteria-carbon-dioxide/ (accessed October 2, 2023).

Smith, William. 2018. "Disruptive Democracy: The Ethics of Direct Action." *Raisons politiques* 69: 13–27.

Snyder, Timothy. 2015a. *Black Earth: The Holocaust as History and Warning*. New York: Tim Duggan Books.

Snyder, Timothy. 2015b. "The Next Genocide." New York Times (September 12).

Snyder, Timothy. 2022. "Ukraine Holds the Future." Foreign Affairs (September 6).

Solnit, Rebecca. 2016. *Hope in the Dark: Untold Histories, Wild Possibilities*. Chicago, IL: Haymarket Books.

Stacey, Jocelyn. 2018. *The Constitution of the Environmental Emergency*. London: Bloomsbury Publishing.

Stanaway, Jeffrey D., Ashkan Afshin, Emmanuela Gakidou, Stephen S. Lim, Amir Radfar, and GBD 2017 Diet Collaborators. 2018. "Global, Regional, and National Comparative Risk Assessment of 84 Behavioural, Environmental and Occupational, and Metabolic Risks or Clusters of Risks for 195 Countries and Territories, 1990–2017: A Systematic Analysis for the Global Burden of Disease Study 2017." *The Lancet* 392: 1923–94.

Starostinetskaya, Anna. 2019. "Fox News Pundits Mock Cory Booker's Vegan Stance with Burgers and Ribs." VegNews (February 14).

Steel, Daniel. 2015. *Philosophy and the Precautionary Principle: Science, Evidence, and Environmental Policy*. Cambridge, UK: Cambridge University Press.

Steel, Daniel, Giulia Belotti, Ross Mittiga, and Kian Mintz-Woo. (unpublished). "A Dynamic Collapse Concept for Climate Change."

Steel, Daniel, C. Tyler DesRoches, and Kian Mintz-Woo. 2022. "Climate Change and the Threat to Civilization." *Proceedings of the National Academy of Sciences* 119: e2210525119.

Steffen, Will, Johan Rockström, Katherine Richardson, Timothy M. Lenton, Carl Folke, Diana Liverman, et al. 2018. "Trajectories of the Earth System in the Anthropocene." *Proceedings of the National Academy of Sciences* 115: 8252–9.

Steinfeld, Henning, Pierre Gerber, Tom Wassenaar, Vincent Castel, Mauricio Rosales, and Cees De Haan. 2006. *Livestock's Long Shadow: Environmental Issues and Options*. Rome: FAO.

Stephenson, Wen. 2017. "Climate Disobedience in the Time of Trump."

Stern, Nicholas. 2007. *The Stern Review: The Economics of Climate Change*. Cambridge, UK: Cambridge University Press.

Stern, Nicholas. 2010a. *A Blueprint for a Safer Planet: How We Can Save the World and Create Prosperity*. New York: Vintage Books USA.

Stern, Nicholas. 2010b. "The Economics of Climate Change." In *Climate Ethics: Essential Readings*, ed. Stephen M. Gardiner, Simon Caney, Dale Jamieson, and Henry Shue. New York: Oxford University Press. 39–76.

Stevenson, Hayley, and John S. Dryzek. 2014. *Democratizing Global Climate Governance*. Cambridge, UK and New York: Cambridge University Press.

Storlazzi, Curt D., Stephen B. Gingerich, Ap van Dongeren, Olivia M. Cheriton, Peter W. Swarzenski, Ellen Quataert, et al. 2018. "Most Atolls will be Uninhabitable by the Mid-21st Century Because of Sea-Level Rise Exacerbating Wave-Driven Flooding." *Science Advances* 4: eaap9741.

Sunstein, Cass R. 2006. "Irreversible and Catastrophic." *Cornell Law Review* 91: 841–98.

Suskind, Ron. 2007. *One Percent Doctrine: Deep inside America's Pursuit of Its Enemies since 9/11*. New York: Simon & Schuster.

Suzman, James. 2017. *Affluence without Abundance: The Disappearing World of the Bushmen*. New York: Bloomsbury USA.

Swanson, Ana. 2015. "An Incredible Image Shows how Powerful Countries are Buying up Much of the World's Land." The Washington Post (May 21).

Sweet, William V., Robert E. Kopp, Christopher P. Weaver, Jayantha Obeysekera, Radley M. Horton, et al. 2017. *Global and Regional Sea Level Rise Scenarios for the United States*. Washington, DC: NOAA Technical Report NOS CO-OPS 083. NOAA/NOS Center for Operational Oceanographic Products and Services.

Tainter, Joseph A. 1989. *The Collapse of Complex Societies*. Cambridge, UK: Cambridge University Press.

Taylor, Matthew. 2022. "More Than 50 Just Stop Oil Protesters in UK Sent to Jail on One Day." The Guardian (September 16).

Thalheimer, Lisa, Moritz P. Schwarz, and Felix Pretis. 2023. "Large Weather and Conflict Effects on Internal Displacement in Somalia with Little Evidence of Feedback onto Conflict." *Global Environmental Change* 79: 102641.

The Fund for Peace. 2021. *Fragile States Index: Annual Report 2021*. Washington, DC: The Fund for Peace.

The Fund for Peace. 2022. "What Do the Colors and Categories in the Index and on the Map Signify?" https://fragilestatesindex.org/frequently-asked-questions/what-do-the-colors-and-categories-in-the-index-and-on-the-map-signify/ (accessed October 19, 2022).

The Syrian Observatory for Human Rights. 2022. "Syrian Revolution 11 Years on | Sohr Documents by Names Nearly 161,000 Civilian Deaths, Including 40,500 Children and Women." https://www.syriahr.com/en/243125/ (accessed October 2, 2023).

The White House. 2022. *National Security Strategy*. Washington, DC: The White House. https://www.whitehouse.gov/wp-content/uploads/2022/10/Biden-Harris-Administrations-National-Security-Strategy-10.2022.pdf.

Thiele, Leslie Paul. 1990. *Friedrich Nietzsche and the Politics of the Soul: A Study of Heroic Individualism*. Vol. 5. Princeton, NJ: Princeton University Press.

Thiele, Leslie Paul. 2013. *Sustainability*. Cambridge, UK and Maldon, MA: Polity Press.

Thompson, Victoria. 1958. "'Golden Rule.'" The Harvard Crimson (May 8).

Tilman, David, Christian Balzer, Jason Hill, and Belinda L. Befort. 2011. "Global Food Demand and the Sustainable Intensification of Agriculture." *Proceedings of the National Academy of Sciences* 108: 20260–4.

Tomlinson, Isobel. 2013. "Doubling Food Production to Feed the 9 Billion: A Critical Perspective on a Key Discourse of Food Security in the UK." *Journal of Rural Studies* 29: 81–90.

Truccone-Borgogno, Santiago. 2023. "Climate Justice and the Duty of Restitution." *Moral Philosophy and Politics* 10: 203–24.

Trump, Donald. 2017. "Statement by President Trump on the Paris Climate Accord." https://trumpwhitehouse.archives.gov/briefings-statements/statement-president-trump-paris-climate-accord/ (accessed October 2, 2023).

Tuck, Richard. 2008. *Free Riding*. Cambridge, MA: Harvard University Press.

Tuck, Richard. 2018. "What Kind of Problem Is Negligibility?: A Response to Melissa Lane." In *Nature, Action and the Future: Political Thought and the Environment*, ed. Katrina Forrester and Sophie Smith. Cambridge, UK: Cambridge University Press. 180–201.

Turnbull, Colin M. 1972. *The Mountain People*. New York: Simon & Schuster.

Union of Concerned Scientists. 2017. "A Brief History of US Fuel Efficiency Standards." December 6 (originally published July 25, 2006). https://www.ucsusa.org/resources/brief-history-us-fuel-efficiency (accessed January 6, 2020).

United Nations. 2009. "Food Production Must Double by 2050 to Meet Demand from World's Growing Population, Innovative Strategies Needed to Combat Hunger, Experts Tell Second Committee." UN Meetings Coverage and Press Releases. October 9. https://press.un.org/en/2009/gaef3242.doc.htm#:~:text=Food%20production%20must%20double%20by%202050%20to%20meet%20the%20demand,a%20panel%20discussion%20on%20%E2%80%9CNew (accessed October 2, 2023).

United Nations. 2021. "Secretary-General Calls Latest IPCC Climate Report 'Code Red for Humanity', Stressing 'Irrefutable' Evidence of Human Influence." UN Meetings Coverage and Press Releases. August 9. https://press.un.org/en/2021/sgsm20847.doc.htm (accessed October 2, 2023).

United Nations Department of Economic and Social Affairs. 2022. *World Population Prospects 2022: Summary of Results*. New York: United Nations Population Division.

United Nations High Commissioner for Refugees. 2022. "Syria Emergency." https://www.unhcr.org/syria-emergency.html (accessed October 19, 2022).

US Department of Defense. 2014a. *2014 Climate Change Adaptation Roadmap*. Alexandria, VA: US Department of Defense.

US Department of Defense. 2014b. "Quadrennial Defense Review 2014." https://www.acq.osd.mil/ncbdp/docs/2014_Quadrennial_Defense_Review.pdf (accessed October 7, 2017).

US Department of Defense. 2015. *National Security Implications of Climate-Related Risks and a Changing Climate*. Washington, DC: US Department of Defense.

van Dijk, Michiel, Tom Morley, Marie Luise Rau, and Yashar Saghai. 2021. "A Meta-Analysis of Projected Global Food Demand and Population at Risk of Hunger for the Period 2010–2050." *Nature Food* 2: 494–501.

van Mill, David. 2021. "Freedom of Speech." *Stanford Encyclopedia of Philosophy* (Spring 2021 edition), ed. Edward N. Zalta. https://plato.stanford.edu/archives/spr2021/entries/freedom-speech/ (accessed October 2, 2023).

Vidal, John. 2008. "Not Guilty: The Greenpeace Activists Who Used Climate Change as a Legal Defence." The Guardian (September 11).

Vohra, Karn, Alina Vodonos, Joel Schwartz, Eloise A. Marais, Melissa P. Sulprizio, and Loretta J. Mickley. 2021. "Global Mortality from Outdoor Fine Particle Pollution Generated by Fossil Fuel Combustion: Results from Geos-Chem." *Environmental Research* 195: 110754.

von Grebmer, Klaus, Jill Bernstein, Miriam Wiemers, Tabea Schiffer, Asja Hanano, Olive Towey, et al. 2021. *Global Hunger Index: Hunger and Food Systems in Conflict Settings*. Dublin and Bonn: Concern Worldwide. https://www.globalhungerindex.org/pdf/en/2021.pdf

von Stein, Jana. 2022. "Democracy, Autocracy, and Everything in Between: How Domestic Institutions Affect Environmental Protection." *British Journal of Political Science* 52: 339–357.

Wagner, Gernot, and Martin Weitzman. 2016. *Climate Shock: The Economic Consequences of a Hotter Planet*. Princeton, NJ: Princeton University Press.

Waldron, Jeremy. 1999. *Law and Disagreement*. Oxford: Clarendon Press.

Walker, K. J. 1988. "The Environmental Crisis: A Critique of Neo-Hobbesian Responses." *Polity* 21: 97–81.

Wall, Tom. 2021. "Wrong to Label Extinction Rebellion as Extremists, Says Home Office Adviser." The Guardian (August 21).

Wallace-Wells, David. 2017. "The Uninhabitable Earth, Annotated Edition." New York Magazine (July 14).

Wallace-Wells, David. 2018. "UN Says Climate Genocide is Coming. It's Actually Worse Than That." New York Magazine (October 10).

Wallimann-Helmer, Ivo. 2015. "Justice for Climate Loss and Damage." *Climatic Change* 133: 469–80.

Walzer, Michael. 2006. *Just and Unjust Wars: A Moral Argument with Historical Illustrations*. New York: Basic Books.

Watts, Jonathan. 2018. "We Have 12 Years to Limit Climate Change Catastrophe, Warns UN." The Guardian (October 8).

Weber, Max. 1991. "Politics as a Vocation." In *From Max Weber: Essays in Sociology*, ed. H. H. Gerth and C. Wright Mills. London: Routledge. 77–128.

Weitzman, Martin. 2012. "GHG Targets as Insurance against Catastrophic Climate Damages." *Journal of Public Economic Theory* 14: 221–44.

Welch, Craig. 2015. "Climate Change Helped Spark Syrian War, Study Says." National Geographic (March 2).

Welchman, Jennifer. 2001. "Is Ecosabotage Civil Disobedience?" *Philosophy & Geography* 4: 97–107.

Wellesley, Laura, Catherine Happer, and Antony Froggatt. 2015. *Changing Climate, Changing Diets: Pathways to Lower Meat Consumption*. London: Chatham House.

Wellman, Christopher H. 1996. "Liberalism, Samaritanism, and Political Legitimacy." *Philosophy & Public Affairs* 25: 211–37.

Wester, Philippus, Arabinda Mishra, Aditi Mukherji, and Arun Bhakta Shrestha, eds. 2019. *The Hindu Kush Himalaya Assessment: Mountains, Climate Change, Sustainability and People*. KathmanduNepal: Springer.

Whiteman, Gail, Chris Hope, and Peter Wadhams. 2013. "Climate Science: Vast Costs of Arctic Change." *Nature* 499: 401–3.

Wiener, Jonathan B. 2016. "Precaution and Climate Change." In *The Oxford Handbook of International Climate Change Law*, ed. Cinnamon P. Carlarne, Kevin R. Gray, and Richard Tarasofsky. New York: Oxford University Press. 163–84.

Wikipedia contributors. 2020. "United States Lighting Energy Policy." Wikipedia, The Free Encyclopedia. Page Version ID: 933737406. https://en.wikipedia.org/wiki/United_States_lighting_energy_policy.

Williams, Bernard. 2005. *In the Beginning Was the Deed: Realism and Moralism in Political Argument*. Princeton, NJ: Princeton University Press.

Williams, Nia, and Laila Kearney. 2016. "Daring US Pipeline Sabotage Spawned by Lobster Boat Coal Protest." Reuters, October 13.

Williston, Byron. 2012. "Climate Change and Radical Hope." *Ethics & the Environment* 17: 165–86.

Williston, Byron. 2015. *The Anthropocene Project: Virtue in the Age of Climate Change*. New York: Oxford University Press.

Wilson, Peter J., Grant McCall, W. R. Geddes, A. K. Mark, John E. Pfeiffer, James B. Boskey, et al. 1975. "More Thoughts on the Ik and Anthropology [and Reply]." *Current Anthropology* 16: 343–58.

Wilson, Tom, and Richard Walton. 2019. *Extremism Rebellion*. London. Policy Exchange.

Wirt, William. 1817. *Sketches of the Life and Character of Patrick Henry*. New York: James Webster.

Wolf, Clark. 2009. "Intergenerational Justice, Human Needs, and Climate Policy." In *Intergenerational Justice*, ed. Axel Gosseries and Lukas H. Meyer. Oxford: Oxford University Press. 347–76.

Wood, Nathan, Rob Lawlor, and Josie Freear. 2023. "Rationing and Climate Change Mitigation*." *Ethics, Policy & Environment* [online] 1–29. https://doi.org/10.1080/21550085.2023.2166342

World Bank. 2012a. "Climate Change Report Warns of Dramatically Warmer World This Century." Press release, November 18. https://www.worldbank.org/en/news/feature/2012/11/18/Climate-change-report-warns-dramatically-warmer-world-this-century (accessed October 9, 2023).

World Bank. 2012b. *Turn Down the Heat*: Why a 4°C Warmer World Must be Avoided. Washington, DC: World Bank.

World Bank. 2022. "The World Bank in DRC." https://www.worldbank.org/en/country/drc/overview#:~:text=DRC%20is%20among%20the%20five,in%20SSA%20lives%20in%20DRC (accessed October 27, 2022).

World Bank Group. 2016. *High and Dry: Climate Change, Water, and the Economy*. Washington, DC: World Bank.

World Food Programme. 2019. "Democratic Republic of the Congo." http://www1.wfp.org/countries/democratic-republic-congo (accessed October 7, 2017).

Wunderling, Nico, Ricarda Winkelmann, Johan Rockström, Sina Loriani, David I. Armstrong McKay, Paul D. L. Ritchie, et al. 2023. "Global Warming Overshoots Increase Risks of Climate Tipping Cascades in a Network Model." *Nature Climate Change* 13: 75–82.

Wündisch, Joachim. 2019. "Partial Loss of Territory Due to Anthropogenic Climate Change: A Theory of Compensating for Losses in Political Self-Determination." *Journal of Applied Philosophy* 36: 313–32.

Wynes, Seth, and Kimberly A. Nicholas. 2017. "The Climate Mitigation Gap: Education and Government Recommendations Miss the Most Effective Individual Actions." *Environmental Research Letters* 12: 074024.

Xu, Chi, Timothy A. Kohler, Timothy M. Lenton, Jens Christian Svenning, and Marten Scheffer. 2020. "Future of the Human Climate Niche." *Proceedings of the National Academy of Sciences* 117: 11350–5.

Xu, Xiaoming, Prateek Sharma, Shijie Shu, Tzu-Shun Lin, Philippe Ciais, Francesco N. Tubiello, et al. 2021. "Global Greenhouse Gas Emissions from Animal-Based Foods are Twice Those of Plant-Based Foods." *Nature Food* 2: 724–32.

Zhao, Qi, Yuming Guo, Tingting Ye, Antonio Gasparrini, Shilu Tong, Ala Overcenco, et al. 2021. "Global, Regional, and National Burden of Mortality Associated with Non-Optimal Ambient Temperatures from 2000 to 2019: A Three-Stage Modelling Study." *The Lancet Planetary Health* 5: e415–25.

Index

For the benefit of digital users, indexed terms that span two pages (e.g., 52–53) may, on occasion, appear on only one of those pages.